D1383954

Springer Laboratory

Springer Laboratory Manuals in Polymer Science

Pasch, Trathnigg: HPLC of Polymers
ISBN: 3-540-61689-6 (hardcover)
ISBN: 3-540-65551-4 (softcover)

Mori, Barth: Size Exclusion Chromatography
ISBN: 3-540-65635-9

Pasch, Schrepp: MALDI-TOF Mass Spectrometry of Synthetic Polymers
ISBN: 3-540-44259-6

Kulicke, Clasen: Viscosimetry of Polymers and Polyelectrolytes
ISBN: 3-540-40760-X

Hatada, Kitayama: NMR Spectroscopy of Polymers
ISBN: 3-540-40220-9

Brummer, R.: Rheology Essentials of Cosmetics and Food Emulsions
ISBN: 3-540-25553-2

Mächtle, W., Börger, L.: Analytical Ultracentrifugation of Polymers
and Nanoparticles
ISBN: 3-540-23432-2

Heinze, T., Liebert, T., Koschella, A.: Esterification of Polysaccharides
ISBN: 3-540-32103-9

Thomas Heinze · Tim Liebert · Andreas Koschella

Esterification of Polysaccharides

With 131 Figures, 105 Tables, and CD-ROM

 Springer

Thomas Heinze
Tim Liebert
Andreas Koschella

Friedrich-Schiller-Universität Jena
Humboldtstraße 10
07743 Jena
Germany
e-mail: thomas.heinze@uni-jena.de
tim.liebert@uni-jena.de
andreas.koschella@uni-jena.de

Library of Congress Control Number: 2006922413

DOI 10.1007/b98412

ISBN-10 3-540-32103-9 **Springer Berlin Heidelberg New York**
ISBN-13 978-3-540-32103-3 **Springer Berlin Heidelberg New York**

e-ISBN 3-540-32112-8

The publisher and the authors accept no legal responsibility for any damage caused by improper use of the instructions and programs contained in this book and the CD-ROM. Although the software has been tested with extreme care, errors in the software cannot be excluded.

Springer is a part of Springer Science+Business Media
springer.com

© Springer-Verlag Berlin Heidelberg 2006
Printed in Germany

The use of general descriptive names, registered names, trademarks, etc. in this publication does not imply, even in the absence of a specific statement, that such names are exempt from the relevant protective laws and regulations and therefore free for general use.

Cover design: *design&production*, Heidelberg, Germany
Typesetting and production: LE-TeX Jelonek, Schmidt & Vöckler GbR, Leipzig, Germany

2/3141 YL 5 4 3 2 1 0 - Printed on acid-free paper

Springer Laboratory Manuals in Polymer Science

Preface

The recent world attention towards renewable and sustainable resources has resulted in many unique and groundbreaking research activities. Polysaccharides, possessing various options for application and use, are by far the most important renewable resources. From the chemist's point of view, the unique structure of polysaccharides combined with many promising properties like hydrophilicity, biocompatibility, biodegradability (at least in the original state), stereoregularity, multichirality, and polyfunctionality, i.e. reactive functional groups (mainly OH−, NH−, and COOH− moieties) that can be modified by various chemical reactions, provide an additional and important argument for their study as a valuable and renewable resource for the future.

Chemical modification of polysaccharides has already proved to be one of the most important paths to develop new products and materials. The objective of this book is to describe esterification of polysaccharides by considering typical synthesis routes, efficient structure characterisation, unconventional polysaccharide esters, and structure-property relationships. Comments about new application areas are also included.

The content of this book originated mainly from the authors' polysaccharide research experience carried out at the Bergische University of Wuppertal, Germany and the Friedrich Schiller University of Jena, Germany. The interaction of the authors with Prof. D. Klemm was a great stimulus to remain active in this fascinating field. In addition, there is increasing interest from industry in the field of polysaccharides that is well documented by the establishment of the Center of Excellence for Polysaccharide Research Jena-Rudolstadt. The aim of the centre is to foster interdisciplinary fundamental research on polysaccharides and their application through active graduate student projects in the fields of carbohydrate chemistry, bioorganic chemistry, and structure analysis.

The authors would like to stress that the knowledge discussed in this book does not represent an endpoint. On the contrary, the information about polysaccharide esters provided here will hopefully encourage scientists in academia and industry to continue the search for and development of new procedures, products, and applications. The authors strongly hope that the polysaccharide ester information highlighted in this book will be helpful both for experts and newcomers to the field.

During the preparation of the book, the members of the Heinze laboratory were very helpful. We thank Dr. Wolfgang Günther for the acquisition of NMR spectra, Dr. Matilde Vieira Nagel for preparing many tables and proofreading the text as well as Stephanie Hornig, Claudia Hänsch, Constance Ißbrücker, and Sarah Köhler for technical assistance. Special thanks go to Prof. Werner-Michael Kulicke, University of Hamburg, who encouraged us to contribute a synthetic topic to the Springer Laboratory series. Dr. Stan Fowler (ES English for Scientists) is gratefully acknowledged for proofreading the manuscript.

The authors would like to express gratitude to Springer for agreeing to publish this book in the Springer Laboratory series. We thank Dr. Marion Hertel of Springer for her conscientious effort.

Jena, February 2006   *Thomas Heinze*
                                                             Tim Liebert
                                                             Andreas Koschella

List of Symbols and Abbreviations

[C$_4$mim]Br	1-N-Butyl-3-methylimidazolium bromide
[C$_4$mim]Cl	1-N-Butyl-3-methylimidazolium chloride
[C$_4$mim]SCN	1-N-Butyl-3-methylimidazolium thiocyanate
Ac	Acetyl
AFM	Atomic force microscope
AGU	Anhydroglucose units
AMIMCl	1-N-Allyl-3-methylimidazolium chloride
APS	Amino propyl silica
Ara$f$	α-L-Arabinofuranosyl
Ara$p$	Arabinopyranosyl
AX	Arabinoxylans
AXU	Anhydroxylose unit
Bu	Butyl
Cadoxen	Cadmiumethylenediamine hydroxide
CDI	N,N'-Carbonyldiimidazole
CI-MS	Chemical ionisation mass spectroscopy
COSY	Correlated spectroscopy
CTFA	Cellulose trifluoroacetate
Cuen	Cupriethylenediamine hydroxide
DB	Degree of branching
DCC	N,N-Dicyclohexylcarbodiimide
DDA	Degree of deacetylation
DEPT	Distortionless enhancement by polarisation transfer
DMAc	N,N-Dimethylacetamide
DMAP	4-N,N-Dimethylaminopyridine
DMF	N,N-Dimethylformamide
DMI	1,3-Dimethyl-2-imidazolidinone
DMSO	Dimethyl sulphoxide
DP	Degree of polymerisation
DS	Degree of substitution
DQF	Double quantum filter
EI-MS	Electron impact ionisation mass spectroscopy
FAB-MS	Fast atom bombardment mass spectroscopy
FACl	Fatty acid chloride
FTIR	Fourier transform infrared spectroscopy

GA	α-D-Glucopyranosyl uronic acid
GalNAc	N-Acetyl-D-galactosamine
Gal$p$	Galactopyranose
Gal$p$N	Galactopyranosylamine
Gal$p$NAc	N-Acetylgalactopyranosylamine
GLC	Gas liquid chromatography
GLC-MS	Gas liquid chromatography-mass spectroscopy
GlcN	D-Glucosamine
GlcNAc	N-Acetyl-D-glucosamine
GlcA	Glucuronic acid
Glc$p$	Glucopyranose
GPC	Gel permeation chromatography
GX	4-O-Methyl-glucuronoxylan
HMBC	Heteronuclear multiple bond correlation
HMPA	Hexamethylphosphor triamide
HMQC	Heteronuclear multiple quantum coherence
HPLC	High-performance liquid chromatography
HSQC	Heteronuclear single quantum correlation
I_c	Crystallinity index
INAPT	Selective version of insensitive nuclei enhanced by polarisation transfer
Maldi-TOF	Matrix assisted laser desorption ionisation time of flight
Man$p$	Mannopyranose
MeGA	4-O-Methyl-α-D-glucopyranosyl uronic acid
MEK	Methylethylketone
MesCl	Methanesulphonic acid chloride
Methyl triflate	Trifluoromethanesulphonic acid methylester
M_w	Mass average molecular mass
n.d.	Not determined
Na dimsyl	Sodium methylsulphinyl
NBS	N-Bromosuccinimide
NIR	Near-infrared
Nitren	Ni(tren)(OH)2[tren=tris(2-aminoethyl)amine]
NMMO	N-Methylmorpholine-N-oxide
NMP	N-Methyl-2-pyrrolidone
NMR	Nuclear magnetic resonance
NOE	Nuclear Overhauser effect
NOESY	Nuclear Overhauser effect spectroscopy
PAHBA	p-Hydroxybenzoic acid hydrazide
PP	4-Pyrrolidinopyridine
Py	Pyridine
RI	Refractive index
RT	Room temperature
RU	Repeating unit
S_N	Nucleophilic substitution

TBA	Tetrabutylammonium
TBAF	Tetrabutylammonium fluoride trihydrate
TBDMS	tert-Butyldimethylsilyl
TDMS	Thexyldimethylsilyl
TEA	Triethylamine
TFA	Trifluoroacetic acid
TFAA	Trifluoroacetic acid anhydride
T_g	Glass transition temperature
THF	Tetrahydrofuran
TMA	Trimethylamine
TMS	Trimethylsilyl
TOCSY	Total correlated spectroscopy
TosCl	p-Toluenesulphonyl chloride
TosOH	p-Toluenesulphonic acid
Trityl	Triphenylmethyl
UV/Vis	Ultraviolet/visible
Xylp	Xylopyranose

Table of Contents

1 Introduction and Objectives

Polysaccharides are unique biopolymers with an enormous structural diversity. Huge amounts of polysaccharides are formed biosynthetically by many organisms including plants, animals, fungi, algae, and microorganisms as storage polymers and structure forming macromolecules due to their extraordinary ability for structure formation by supramolecular interactions of variable types. In addition, polysaccharides are increasingly recognised as key substances in biotransformation processes regarding, e.g., activity and selectivity. Although the naturally occurring polysaccharides are already outstanding, chemical modification can improve the given features and can even be used to tailor advanced materials.

Etherification and esterification of polysaccharides represent the most versatile transformations as they provide easy access to a variety of bio-based materials with valuable properties. In particular, state-of-the-art esterification can yield a broad spectrum of polysaccharide derivatives, as discussed in the frame of this book from a practical point of view but are currently only used under lab-scale conditions. In contrast, simple esterification of the most abundant polysaccharides cellulose and starch are commercially accepted procedures. Nevertheless, it is the author's intention to review classical concepts of esterification, such as conversions of cellulose to carboxylic acid esters of C_2 to C_4 acids including mixed derivatives of phthalic acid and cellulose nitrate, which are produced in large quantities. These commercial paths of polysaccharide esterification are carried out exclusively under heterogeneous conditions, at least at the beginning of the conversion. The majority of cellulose acetate (about 900 000 t per year) is based on a route that includes the dissolution of the products formed [1–3].

Research and development offers new opportunities for the synthesis of polysaccharide esters resulting from:

- New reagents (ring opening, transesterification), enzymatic acylation and in situ activation of carboxylic acids
- Homogeneous reaction paths, i.e., starting with a dissolved polysaccharide and new reaction media
- Regioselective esterification applying protecting-group techniques and protecting-group-free methods exploiting the superstructural features of the polysaccharides as well as enzymatically catalysed procedures

With regard to structure characterisation on the molecular level most important are NMR spectroscopic techniques including specific sample preparation. Having

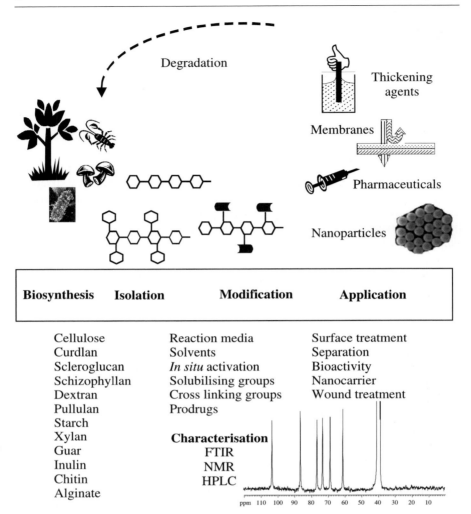

Biosynthesis	Isolation	Modification	Application
Cellulose		Reaction media	Surface treatment
Curdlan		Solvents	Separation
Scleroglucan		*In situ* activation	Bioactivity
Schizophyllan		Solubilising groups	Nanocarrier
Dextran		Cross linking groups	Wound treatment
Pullulan		Prodrugs	
Starch			
Xylan		**Characterisation**	
Guar		FTIR	
Inulin		NMR	
Chitin		HPLC	
Alginate			

been extensively involved in polysaccharide research, we would like to stress that a clear description of structure–property relationships is conveniently accessible not only for the commercial derivatives, but also for products of improved or even new features by this technique.

The combination of new esterification techniques, comprehensive structure characterisation and detailed structure–property relationships is the key for nanoscience and nanotechnology, smart and responsive materials with polysaccharides and also opens new applications in the field of biosensors, selective separation, bioengineering and pharmaceutics.

The objective of the book is not to supplement or replace any of the several review articles and books in the field of polysaccharide chemistry and in particular esterification, but rather to describe the important features of typical synthetic routes, efficient structure characterisation, and unconventional polysaccharide esters including structure–property relationships. Additionally, comments about selected new application areas are included. The methods of modification and analysis described are mainly focused on glucans because they represent a large part of naturally occurring polysaccharides. Moreover, glucans are structurally most uniform. In contrast, polysaccharides consisting of various monosaccharides and substructures, e.g., galactomannans, or algal polysaccharides exhibit a broad diversity in properties caused by a large number of irreproducible factors. Thus, the features of the algal polysaccharides vary extensively depending on the seaweed species, the part of the plant the alginate was extracted from, and climatic conditions at the time of growth [4]. Modification multiplies the structural and property broadness, and is therefore of limited relevance for these polymers up to now. Structure analysis is hardly achievable. Because most analytical strategies and synthesis paths are adapted from the conversion of glucans, more complex polymers will only be discussed if specific treatment is applied, e.g., esterification of carboxylic acid moieties of alginates [5]. Among the broad variety of these complex polysaccharides, the most important galactomannan guar gum, the algal polysaccharide alginate, the aminoglucane chitin, the hemicellulose xylan, and the fructan inulin are discussed to demonstrate the specifics of these polymers.

Although recently the chemical (ring opening polymerisation) and enzymatic synthesis of polysaccharides and polysaccharide derivatives was experimentally achieved (up to now rather low DP values of maximum 40 have been obtained) the polymeranalogous modification of polysaccharides isolated from natural sources is the most important route to new products today and will continue to be the most important in the foreseeable future. Consequently, polymeranalogous reactions are discussed exclusively. It should be pointed out that not necessarily a strictly polymeranalogous reaction (no change in DP) is required. On the contrary, a certain degradation prior or during the reaction may be a desired goal.

We hope this book fills a gap between various aspects of polysaccharide research concerning biosynthesis and isolation, on one hand, and material science, on the other hand. It is hoped that this book will be accepted by the scientific community as a means to stimulate scientists from different fields to use the chemical modification of polysaccharides as basis for innovative ideas and new experimental pathways.

2 Structure of Polysaccharides

2.1 Structural Features

There is a wide range of naturally occurring polysaccharides derived from plants, microorganisms, fungi, marine organisms and animals possessing magnificent structural diversity and functional versatility. In Table 2.1, polysaccharides most commonly used for polymeranalogous reactions are summarised according to chemical structures. These include glucans (1–8), fructans (11), aminodeoxy glucans (12, 13), and polysaccharides with uronic acid units (14).

Table 2.1. Structures of polysaccharides of different origin

Polysaccharide Type		Source	Structure	Reference
Cellulose	1	Plants	β-(1→4)-D-glucose	[6]
Curdlan	2	Bacteria	β-(1→3)-D-glucose	[7]
Scleroglucan	3	Fungi	β-(1→3)-D-glucose main chain, β-(1→6)-D-glucose branches	[8]
Schizophyllan	4	Fungi	β-(1→3)-D-glucose main chain, D-glucose branches	[9, 10]
Dextran	5	Bacteria	α-(1→6)-D-glucose main chain	[11]
Pullulan	6	Fungi	α-(1→6) linked maltotriosyl units	[12]
Starch		Plants		[13]
Amylose	7		α-(1→4)-D-glucose	
Amylopectin	8		α-(1→4)- and α-(1→6)-D-glucose	
Xylan	9	Plants	β-(1→4)-D-xylose main chain	[14]
Guar	10	Plants	β-(1→4)-D-mannose main chain, D-galactose branches	[15]
Inulin	11	Plants	β-(1→2)-fructofuranose	[16]
Chitin	12	Animals	β-(1→4)-D-(N-acetyl)glucosamine	[17]
Chitosan	13		β-(1→4)-D-glucosamine	
Alginate	14	Algae	α-(1→4)-L-guluronic acid β-(1→4)-D-mannuronic acid	[18]

The common motifs are primary and secondary OH groups and carboxylic acid moieties, accessible to esterification, and NH_2 groups for conversion to amides. In addition, comprehensive reviews about the molecular, supramolecular and morphological structures of the polysaccharides are available [9, 19–23].

2.1.1 Cellulose

Cellulose, the most abundant organic compound, is a linear homopolymer composed of D-glucopyranose units (so-called anhydroglucose units) that are linked together by β-(1→4) glycosidic bonds (Fig. 2.1). Although cellulose possesses a unique and simple molecular structure, very complex supramolecular structures can be formed, which show a rather remarkable influence on properties such as reactivity during chemical modification. The complexity of the different structural levels of cellulose, i.e. the molecular, supramolecular and morphological, is well studied [24]. The polymer is insoluble in water, even at a rather low DP of 30, and in common organic solvents, resulting from the very strong hydrogen bond network formed by the hydroxyl groups and the ring and bridge oxygen atoms both within and between the polymer chains. The ordered hydrogen bond systems form various types of supramolecular semicrystalline structures. This hydrogen bonding has a strong influence on the whole chemical behaviour of cellulose [25, 26].

To dissolve the polymer, various complex solvent mixtures have been evaluated and are most often employed in esterification reactions such as DMAc/LiCl and DMSO/TBAF. A well-resolved ^{13}C NMR spectrum of the polymer dissolved in DMSO-d_6/TBAF, including the assignment of the 6 carbon atoms, is shown in Fig. 2.1 [27].

The carbon atoms of position 2, 3 and 6 possess hydroxyl groups that undergo standard reactions known for primary and secondary OH moieties. Cellulose of various DP values is available, depending on the source and pre-treatment. Native cotton possesses values up to 12 000 while the DP of scoured and bleached cotton linters ranges from 800 to 1800 and of wood pulp (dissolving pulp) from 600 to 1200.

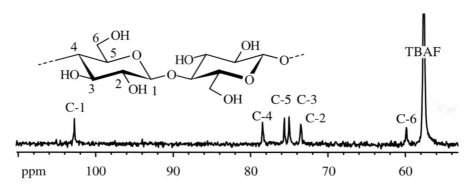

Fig. 2.1. ^{13}C NMR spectrum of cellulose dissolved in DMSO-d_6/TBAF (reproduced with permission from [27], copyright Wiley VCH)

Table 2.2. Carbohydrate composition, DP, and crystallinity of commercially available celluloses

Sample	Producer	Carbohydrate composition (%)			DP	Crystallinity (%)
		Glucose	Mannose	Xylose		
Avicel	Fluka	100.0	–	–	280	61
Sulphate pulp V-60	Buckeye	95.3	1.6	3.1	800	54
Sulphate pulp A-6	Buckeye	96.0	1.8	2.2	2000	52
Sulphite pulp 5-V-5	Borregaard	95.5	2.0	2.5	800	54
Linters	Buckeye	100.0	–	–	1470	63

Table 2.2 gives some examples of cellulose with a high variety of DP values useful for chemical modification. Another approach to pure cellulose is the laboratory-scale synthesis of the polymer by *Acetobacter xylinum* and *Acanthamoeba castellani* [28], which circumvents problems associated with the extraction of cellulose.

2.1.2 β-(1→3)-Glucans

There are a number of structural variations within the class of polysaccharides classified as β-(1→3)-glucans. The group of β-(1→3, 1→6) linked glucans has been shown to stimulate and enhance the human immune system.

Although polysaccharides of the curdlan type are present in a variety of living organisms including fungi, yeasts, algae, bacteria and higher plants, until now only bacteria belonging to the *Alcaligenes* and *Agrobacterium* genera have been reported to produce the linear homopolymer. Curdlan formed by bacteria including *Agrobacterium biovar* and *Alcaligenes faecalis* is a homopolymer of β-(1→3)-D-glucose, determined by both chemical and enzymatic analysis (Fig. 2.2, [29]). Thus, this β-glucan is unbranched. The DP is approximately 450 and the polymer is soluble in both DMSO and dilute aqueous NaOH. About 700 t of the polysaccharide are commercially produced in Japan annually.

Scleroglucan is a neutral homopolysaccharide consisting of linear β-(1→3) linked D-glucose, which contains a β-(1→6) linked D-glucose at every third repeating unit of the main chain on average (Fig. 2.2, [8]). The polysaccharide is soluble in water and DMSO. Scleroglucan is formed extracellularly by *Sclerotium glucanicum* and other species of *Sclerotium*. The polysaccharide schizophyllan

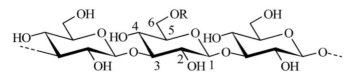

Fig. 2.2. Chemical structure of β-(1→3)-glucans: curdlan (R = H), scleroglucan (R = β-D-glucopyranosyl moiety)

synthesised by *Schizophyllum commune* possesses the same primary structure as scleroglucan [30].

Scleroglucan, schizophyllan and curdlan have found some attention within the context of chemical modification.

2.1.3 Dextran

Dextran, produced by numerous strains of bacteria (*Leuconostoc* and *Strepto-coccus*), is a family of neutral polysaccharides consisting of a α-(1→6) linked D-glucose main chain with varying proportions of linkages and branches, depending on the bacteria used. The α-(1→6) linkages in dextran may vary from 97 to 50% of total glycosidic bonds. The balance represents α-(1→2), α-(1→3), and α-(1→6) linkages usually bound as branches [31]. The commercially applied single strain of *Leuconostoc mesenteroides* NRRL B-512F produces a dextran extracellularly (Fig. 2.3) that is linked predominately by α-(1→6) glycosidic bonds with a relatively low level (∼5%) of randomly distributed α-(1→3) branched linkages [32]. The majority of side chains (branches) contain one to two glucose units. The dextran of this structure is generally soluble in water and other solvents (formamide, glycerol). The commercial production carried out by various companies is estimated to be ca. 2000 t/year worldwide [33].

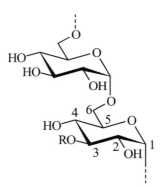

Fig. 2.3. Structure of dextran obtained from *Leuconostoc mesenteroides* NRRL B-512F. R = predominately H and 5% glucose or α-(1→6) linked glucopyranosyl-α-D-glucopyranoside

2.1.4 Pullulan

Pullulan is a water-soluble, neutral polysaccharide formed extracellularly by certain strains of the polymorphic fungus *Aureobasidium pullulans*. It is now widely accepted that pullulan is a linear polymer with maltotriosyl repeating units joined by α-(1→6) linkages [12, 34]. The maltotriosyl units consist of α-(1→4) linked D-glucose (Fig. 2.4). Consequently, the molecular structure of pullulan is intermediate between amylose and dextran because it contains both types of glycosidic bonds in one polymer.

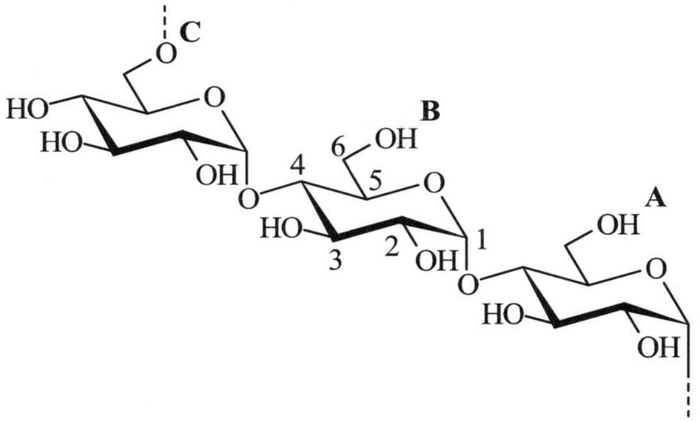

Fig. 2.4. Structure of pullulan

The polysaccharide possesses hydroxyl groups at position 2, 3 and 4 of different reactivity (Fig. 2.4). The structure of pullulan has been analysed by ^{13}C and ^{1}H NMR spectroscopic studies using D_2O or DMSO-d_6 as solvents [35]. The repeating unit linked by α-(1→6) bond shows a greater motional freedom than the units connected by α(1→4), which may influence the functionalisation pattern obtained by chemical modification in particular homogeneously in dilute solution.

2.1.5 Starch

Starch consists of two primary polymers containing D-glucose, namely the linear α-(1→4) linked amylose and the amylopectin that is composed of α-(1→4) linked D-glucose and α-(1→6) linked branches (Fig. 2.5). The molecular mass of amylose is in the range 10^5–10^6, while amylopectin shows significantly higher values of 10^7–10^8 [13]. Amylose and amylopectin occur in varying ratios depending on the plant species (Table 2.3).

Table 2.3. Typical starch materials, their composition, and suppliers

Starch type	Amylose content (%)	Supplier	Contact
Hylon VII	70	National starch	www.nationalstarch.com
Amioca powder	1	National starch	www.nationalstarch.com
Potato starch	28	Emsland Stärke	www.emsland-staerke.de
Waxy maize starch	1	Cerestar	www.cerestar.com

2.1.6 Hemicelluloses

Hemicelluloses are among the most abundant polysaccharides in the world, since they constitute 20–30% of the total bulk of annual and perennial plants. According to the classical definition, hemicelluloses are cell wall polysaccharides that are

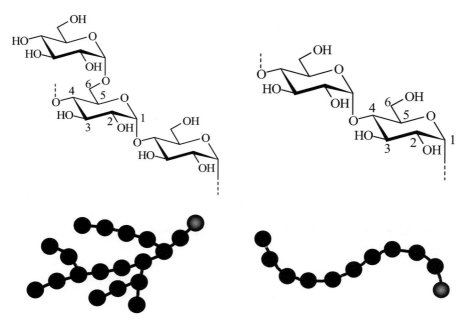

Fig. 2.5. Structures of amylopectin (*left*) and amylose (*right*) and schematic representation of the branching pattern

extractable by aqueous alkaline media. Hemicelluloses possess a broad structural diversity [36]. Xylans, mannans and galactans are present in wood.

Xylans

The xylan-type polysaccharides, the most frequently occurring hemicelluloses, are known to occur in several structural varieties in terrestrial plants and algae, and even in different plant tissues within one plant (Fig. 2.6) [14].

Xylans of higher plants possess β-(1→4) linked Xyl$p$ units as the backbone, usually substituted with sugar units and O-acetyl groups. In the wood of deciduous trees, only the GX type (Fig. 2.6a) was found to be present, which contains single side chains of 2-linked MeGA units. The xylose to MeGA ratios of GX isolated from different hardwoods vary in the range 4–16:1.

Arabino(glucurono)xylan types containing single side chains of 2-O-linked α-D-glucopyranosyl uronic acid unit and/or its 4-O-methyl derivative (MeGA) and 3-linked Ara$f$ units (Fig. 2.6b) are typical of softwoods and the lignified tissues of grasses and annual plants. Neutral arabinoxylans with Xyl$p$ residues substituted at position 3 and/or at both positions 2 and 3 of Xyl$p$ by α-L-Ara$f$ units represent the main xylan component of cereal grains.

Highly branched water-soluble AX (Fig. 2.6c), differing in frequency and distribution of mono- and disubstituted Xyl$p$ residues, are present in the endospermic as well as pericarp tissues. The DP of xylans varies from approximately 100 to 200.

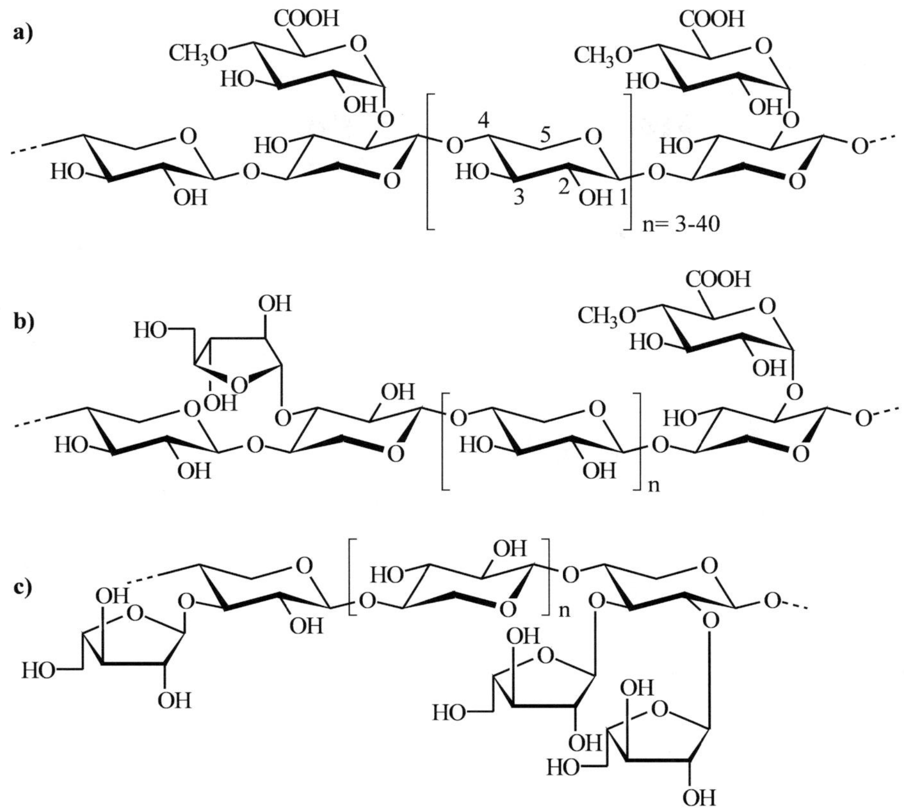

Fig. 2.6. Structures of (**a**) 4-O-methylglucuronoxylan, (**b**) arabino-(glucurono)-xylan, and (**c**) arabinoxylan

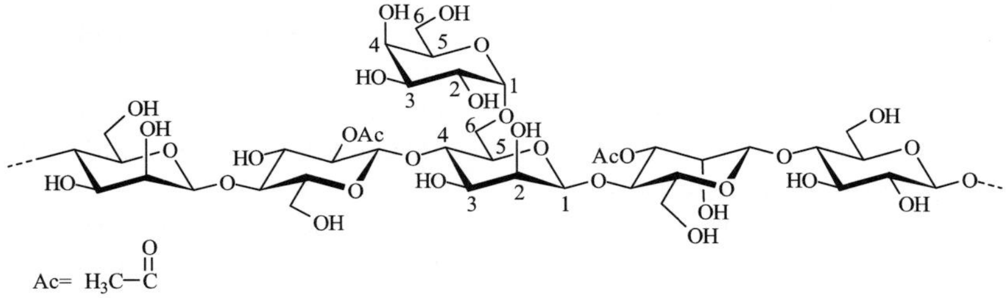

$$\rightarrow4\text{-}\beta\text{-D-Man}p\text{-}1\rightarrow4\text{-}\beta\text{-D-Glc}p\text{-}1\rightarrow4\text{-}\beta\text{-D-Man}p\text{-}1\rightarrow4\text{-}\beta\text{-D-Man}p\text{-}1\rightarrow4\text{-}\beta\text{-D-Glc}p$$

Fig. 2.7. Structure of a softwood glucomannan

Mannans

In coniferous trees, mannans containing mannose, glucose and galactose acetylated to various extents are found. A typical glucomannan from softwood is depicted in Fig. 2.7.

2.1.7 Guar

Guar is a typical example of plant gums that form viscous aqueous solutions. Guar gum is a seed extract containing mannose with galactose branches every second unit. In the galactomannan, the mannose is β-(1→4) connected, while the D-galactose is attached via α-(1→6) links (Fig. 2.8). The sugar ratio is approximately 1.8:1 and irregularities in the pattern of side groups are well known [15]. Guar, isolated from natural sources, can have molecular mass up to 2 000 000 g/mol.

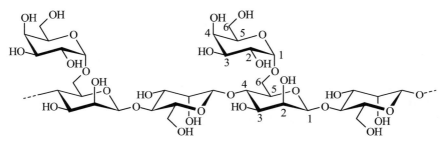

Fig. 2.8. Structure of guar

2.1.8 Inulin

Inulin is an example of so-called fructans, polysaccharides that are widely spread in the vegetable kingdom. Inulin consists mainly of β-(1→2) linked fructofuranose units. A starting glucose moiety is present. The DP of plant inulin varies according to the plant species but is usually rather low. The most important sources are chicory (*Cichorium intybus*), dahlia (*Dahlia pinuata* Cav.) and Jerusalem artichoke (*Helianthus tuberosus*). The average DP is 10–14, 20 and 6 respectively. Inulin may be slightly branched. The amount of β-(2→6) branches in inulin from chicory and dahlia is 1–2 and 4–5% respectively. In contrast, bacterial inulin has high DP values ranging from 10 000 to 100 000, and is additionally highly branched [16,37] (Fig. 2.9).

2.1.9 Chitin and Chitosan

Chitin is widely distributed amongst living organisms, with crabs, prawns, shrimps and freshwater crayfish being most commercially important. Although crustaceans are harvested for human food purposes, they are also the source of chitin, which

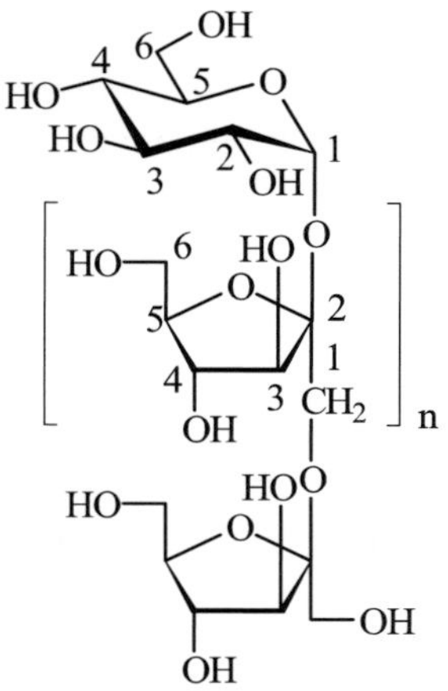

Fig. 2.9. Structure of inulin

is isolated by treatment with aqueous NaOH. Chitin consists of β-(1→4) linked GlcNAc whereas chitosan is the corresponding polysaccharide of GlcN. However, both polysaccharides do not show the ideal structure of a homopolymer since they contain varying fractions of GlcNAc and GlcN residues (Fig. 2.10). To distinguish between the two, it is most appropriate to use solubility in 1% aqueous solution of acetic acid. Chitin containing about 40% of GlcNAc moieties is insoluble while the soluble polysaccharide is named chitosan [38].

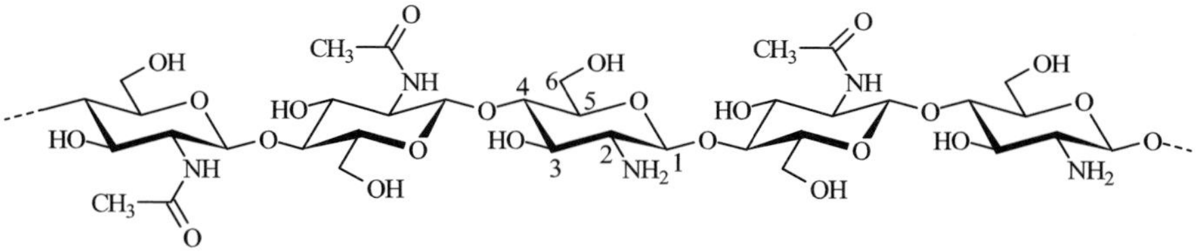

Fig. 2.10. Structure of chitin consisting of N-acetylglucosamin and glucosamin units (DDA 40%)

Table 2.4. Selected companies offering chitin and chitosan (adapted from [38])

Company	Contact
Henkel KGaA, Düsseldorf, Germany	www.bioprawns.no
Genis hf, Iceland	www.genis.is
Kate International	www.kateinternational.com
Kitto Life Co., Seoul, Korea	www.kittolife.co.kr
Micromod GmbH, Germany	www.micromod.de
Primex Ingredients ASA, Norway	www.primex.no
Pronova, Norway	www.pronova.com

The DP of chitin is in the range of 5000 up to 10 000. Owing to hydrogen bonds, chitin occurs in three different polymorphic forms that differ in the orientation of the polymer chains. Mostly, the thermodynamically stable α-chitin and the metastable β-chitin occur. There are various suppliers for pure chitin worldwide (Table 2.4).

2.1.10 Alginates

Alginate is a gelling polysaccharide found in high abundance in brown seaweed. Being a family of unbranched copolymers, the primary structure of alginates varies greatly and depends on the alga species as well as on seasonal and growth conditions. The three commercially most important genera are *Macrocystis*, *Laminaria* and *Ascophyllum* [18]. The repeating units of alginates are α-L-guluronic acid and β-D-mannuronic acid linked by 1→4 glycosidic bonds of varying composition and sequence. The polymer chain contains blocks of guluronic acid and mannuronic acid as well as alternating sequences (Fig. 2.11).

Alginates with a more uniform structure containing preferably mannuronic acid (up to 100%) are found in bacteria [39]. In addition, alginates of high guluronic acid content can be prepared by chemical treatment of alginates and fractionation. By an enzymatic in vitro treatment of alginates with mannuronan C-5 epimerase, the guluronic acid content of the polysaccharide can be increased by epimerization of the C-5 centre of α-L-guluronic acid to give β-D-mannuronic acid.

In view of the fact that the structural features of the polysaccharides discussed above may change due to, for example, seasonal conditions, comprehensive analysis of the specific biopolymer is recommended as discussed in the next chapter.

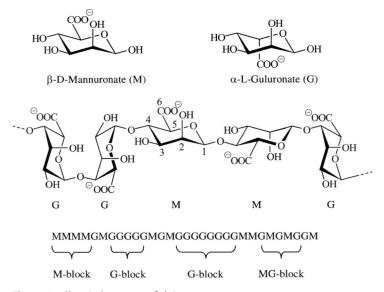

Fig. 2.11. Chemical structure of alginate

3 Analysis of Polysaccharide Structures

A broad variety of specific methods for the structure analysis of polysaccharides, their interaction with different compounds such as solvents or inorganic salts, and the superstructures both in solid state and in solution have been established. An overview of methods and results for the superstructural behaviour of polysaccharides is given in [19]. The aim of this chapter is to present a review of the techniques that can be performed on commercially available equipment to elucidate the primary structure of polysaccharides.

It is an essential prerequisite to analyse the polysaccharides before modification as comprehensively as possible to monitor all types of structural changes of the polymer backbone during the conversion to a derivative. One should always keep in mind that purification beyond the removal of low molecular mass impurities is not reasonable. The basic RU of the polysaccharides described in the book are given in Chap. 2. Nevertheless, analysis of the polysaccharide in question is always recommended because the chemical structure, including branching, sequences of sugar units, oxidised moieties in the chain (e.g. aldehyde-, keto-, and carboxylic groups in polyglucans), and the residual amount of naturally occurring impurities vary for a given type of polysaccharide, especially for fungal and plant polymers, and may significantly influence the properties and reactivity.

A number of basic chemical methods have been developed for the structure analysis and the determination of the purity of polysaccharides. Most of these chemical analyses are colour reactions, which can be quantified by UV/Vis spectroscopy. A list of methods and the features determined is shown in Table 3.1.

In addition, for ionic polymers such as alginates or chitosan salt, titration can be exploited to obtain information about the number of functional groups within the polymer. Linear potentiometric titration is used for the determination of free amino functions in chitinous materials [45].

A value that should be analysed carefully before conversion of a polysaccharide to an ester is the amount of absorbed water in the starting polymer. This is possible by thermogravimetry or by amperometric titration with Karl Fischer reagent after water extraction. In the case of cellulose extraction, the most suitable extractants are DMF, acetonitrile and isobutanol [46].

Table 3.1. Summary of chemical methods used for structure determination of polysaccharides

Test	Method	λ_{max} (nm)	Detected structure	Ref.
Anthron	Anthron in H_2SO_4	620	Free and bound hexose on polysaccharides Blue: hexose, Green: other sugars	[41]
Oricinol	Oricinol in EtOH and $FeCl_3$ in HCl	665	Free and bound pentoses on polysaccharides Green to blue: pentose produce green to blue coloration	[40]
Phenol/ H_2SO_4	Phenol and H_2SO_4	485	Free and bound sugars in soluble and insoluble polysaccharides	[41]
Biphenylol	Hydroxybiphenylol in NaOH and borax in H_2SO_4	520	Free and bound uronic acid on polysaccharides (red to blue)	[42]
Cystein/ H_2SO_4	Cystein-HCl in H_2O and H_2SO_4	380, 396, 427	Free and bound 6-desoxyhexose on polysaccharides	[40]
PAHBAH	PAHBAH in HCl and NaOH	410	Reducing sugars	[43]
Updegraff	$AcOH/H_2O/$ HNO_3 (8:2:1)		Cellulose	[44]

3.1 Optical Spectroscopy

Besides the above-mentioned analysis of polysaccharides with UV/Vis spectroscopy after chemical treatment, optical spectroscopy is used for some semiquantitative methods for the determination of the amount of functional groups (NH-CO-CH_3, $COOH$). The DDA value in chitinous material can be determined via UV/Vis measurements at 210 nm after dissolution in 85% phosphoric acid under thermal-controlled sonication [47].

Optical spectroscopy can be used to determine the conformation of structural features of pure polysaccharides and to easily monitor structural changes during modification. FTIR spectroscopy yields "fingerprint" spectra usable as structural evidence. The most common way for FTIR measurements is the preparation of KBr pellets. To obtain well-resolved spectra, it is necessary to apply a ball mill to guarantee homogeneous mixtures of KBr and the macromolecule. Usually, samples containing about 1–2% (w/w) polymer are prepared. Common "non-polymer" signals observed by means of FTIR spectroscopy are adsorbed water at about 1630–1640 cm^{-1} and CO_2 at about 2340–2350 cm^{-1}. A number of FTIR spectra obtained for the glucanes cellulose, starch, dextran and scleroglucan are shown in Fig. 3.1. The general assignment is given in Table 3.2.

Alginates show additional signals for the $C=O$ moiety of the carboxylate at 1620–1630 and 1410–1420 cm^{-1} or at 1730 cm^{-1}, if the alginate is transferred to

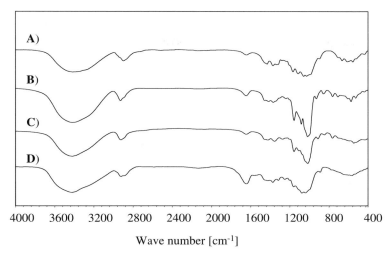

Wave number [cm^{-1}]

Fig. 3.1. FTIR spectra of the glucanes **A** cellulose, **B** starch, **C** dextran and **D** scleroglucan

Table 3.2. General assignment of FTIR spectra of polysaccharides (adapted from [48])

Wave number (cm^{-1})	Assignment
3450–3570	OH stretch, intramolecular H-bridge between the OH groups
3200–3400	OH stretch, intermolecular H-bridge between the OH groups
2933–2981	CH$_2$ antisymmetric stretch
2850–2904	CH$_2$ symmetric stretch
1725–1730	C=O stretch from acetyl- or COOH groups
1635	Adsorption of water
1455–1470	CH$_2$ symmetric ring stretch at pyrane ring; OH in-plane deformation
1416–1430	CH$_2$ scissors vibration
1374–1375	CH deformation
1335–1336	OH in-plane deformation
1315–1317	CH$_2$ tip vibration
1277–1282	CH deformation
1225–1235	OH in-plane deformation, also in COOH groups
1200–1205	OH in-plane deformation
1125–1162	C–O–C antisymmetric stretch
1107–1110	Ring antisymmetric stretch
1015–1060	C–O stretch
985–996	C–O stretch
925–930	Pyran ring stretch
892–895	C-anomeric groups stretch, C1–H-deformation; ring stretch
800	Pyran ring stretch

the acid form (see Table 3.3). A semi-quantitative determination of the ratios of the two sugar residues in the polymer is possible by comparing the peak areas in spectra at 808 cm^{-1} for β-D-mannuronic acid and 887 cm^{-1} for α-L-guluronic acid [49].

Table 3.3. Characteristic bands for alginates and galactomannanes in FTIR spectra (v in cm^{-1}, adapted from [50])

Alginic acid	Sodium alginate	Mannan	Galactomannan
1730	–	–	
1610	1605	1630	–
–	1405	–	1635
920	935	935	–
870	880	870	–
805	810	805	860
–	–	760	805
660	–	–	760 (weak)
–	–	600	

Galactomannans show characteristic FTIR signals at 870 cm^{-1} and in the range 805–820 cm^{-1} [50]. Via FTIR spectroscopy, the glucans and mannans isolated from yeast, e.g. *Saccharomyces cerevisiae*, can be analysed for composition using a characteristic mannan signal at 805 cm^{-1} [51]. The amount of xylan in ligno-celluloses can be determined by evaluation of the carboxyl bands at 1736 cm^{-1} in deconvoluted FTIR spectra [52].

Chitin possesses typical signals at 1650, 1550 and 973 cm^{-1}, which are the amide I, II and III bands respectively. A sharp signal appears at 1378 cm^{-1}, caused by the CH$_3$ symmetrical deformation. The amount of N-acetylation can be estimated from the signal areas of the bands at 1655 and 3450 cm^{-1}, according to the following equation [53].

$$\% N(\text{acetyl}) = \frac{A_{1655}}{A_{3450}} \cdot 115 \tag{3.1}$$

3.2 NMR Spectroscopy

The most powerful tool for polysaccharide analysis is NMR spectroscopy. For the majority of these biopolymers, well-resolved ^{1}H-, ^{13}C- and two-dimensional NMR spectra can be acquired from solutions of the intact polymers in DMSO-d_6, in D$_2$O, and in other deuterated solvents (solubility, see Table 5.1 in Chap. 5, chemical shifts of the solvents, see Table 8.2 in Chap. 8). Restrictions exist only for cellulose and chitin, which are not easily soluble, and for guar gum, alginates and scleroglucanes, which lead to highly viscose solutions at low concentrations, yielding badly resolved spectra. In the case of cellulose and chitin, the application of solid state NMR spectroscopy or the use of specific solvents are necessary. For the other polysaccharides, careful acidic degradation is recommended (see Sect. 3.2.4).

3.2.1 ^{13}C NMR Spectroscopy

For ^{13}C NMR spectroscopy, solutions containing 8 to 10% (w/w) polymer should be used, if the viscosity of the solutions permits. In order to circumvent problems due to high viscosity, polymer degradation by means of acidic hydrolysis (see Table 3.17) and ultrasonic degradation may be applied [54]. The measurements should be carried out at elevated temperature. If the polysaccharides do not dissolve sufficiently in DMSO, brief heating to 80 °C can be helpful, or the addition of small amounts of LiCl. The solvent applied has an influence on the chemical shifts of the signals. Measurements in D_2O commonly lead to a downfield shift (higher ppm values) in the range of 1–2 ppm. In Table 3.4, an overview of relevant chemical shifts and corresponding carbon atoms for ^{13}C NMR signals of polysaccharides is given.

Table 3.4. General overview of chemical shifts and the corresponding carbon atoms for ^{13}C NMR signals of polysaccharides

C atom (moiety)	Chemical shift (ppm)
C-1 (β)	103–106
C-1 (α)	98–103
Involved in glycosidic bond, non-anomeric C	77–87
C-2 to C-5	65–75
CH_2OH	60–62
C=O (carboxylic acid moieties)	175–180
COOH	170–175
O–CH_3	55–61
O–(C=O)CH_3	20–23
CH_3	15–18

A detailed assignment of a variety of polysaccharides described in this book is shown in Table 3.5. In general, the signals for the carbon atoms of the glycosidic linkages (C-1) indicate the type of anomer the RU represents. Thus, polymers built up from β-anomers show a C-1 signal at about 103 ppm, e.g. curdlan (Fig. 3.2) or cellulose (Fig. 2.1).

Polymers consisting of α-anomers yield signals at approximately 98–100 ppm for C-1. The involvement of C-3, C-4 or C-6 in a glycosidic linkage is usually

Table 3.5. Detailed assignment of ^{13}C NMR shifts of polysaccharides

Polysaccharide	Chemical shift (ppm)						Ref.
	C-1	C-2	C-3	C-4	C-5	C-6	
Scleroglucan (DMSO-d_6)	104.6	74.1	88.0	70.0	76.5	70.1	[55]
Galactomannan–galactose (D$_2$O)	101.5	71.2	72.0	72.2	73.8	63.9	[56]
Pullulan (1→4)-(1→6)-(1→4) Glc (D$_2$O)	98.8	72.1	74.2	78.7	72.1	61.8	[57]
Curdlan (DMSO-d_6)	103.0	72.8	86.2	68.4	76.3	60.8	[51]
Dextran (D$_2$O)	98.1	71.8	73.3	70.0	70.3	66.0	[58]
Starch (DMSO-d_6)	99.9	71.5	73.1	78.6	69.9	60.4	
Alginate	102.2	67.4	71.8	80.3	69.9	177.1	[59]
	103.9	73.3	74.3	82.6	78.9	177.6	
Xylan	101.5	72.4	73.8	75.3	63.0	–	
Inulin	62.1	103.8	78.0	75.4	82.5	62.4	
Cellulose in DMAc/LiCl	103.9	74.9	76.6	79.8	76.6	60.6	
Chitosan in CD$_3$COOD	100.5	58.7	73.0	794	77.7	62.9	

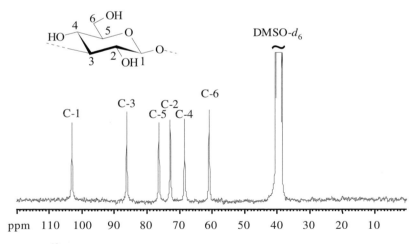

Fig. 3.2. ^{13}C NMR spectrum of curdlan from *Alcaligenes faecalis* in DMSO-d_6

indicated by a downfield shift (towards higher ppm values) of the corresponding ^{13}C signal in the range of 7–11 ppm. The C-4 signal of (1→4)-glucans such as cellulose and starch is at approximately 78 ppm and the C-6 signal at approximately 60 ppm. In contrast, the C-4 peak of the (1→6)-glucan dextran is at 70 ppm and the C-6 signal at 67 ppm (Fig. 3.3, [58]).

A comparable assignment (Table 3.5) is found for polysaccharides consisting of sugars other than glucose. The spectra of inulin (fructan) and xylan (consisting mainly of xylose) show signals for the glycosidic C-1 at 101–104 ppm, for the CH_2OH moiety at 62–63 ppm, and for the secondary C-atoms in the range 72–83 ppm. In the case of inulin, two peaks at 62.1 and 62.4 ppm are observed for the primary carbon of the fructose and the glucose units. The carbonyl signal of the carboxylate moieties in xylan is usually not visible for purified samples, and thus the ^{13}C NMR spectrum shows five sharp peaks.

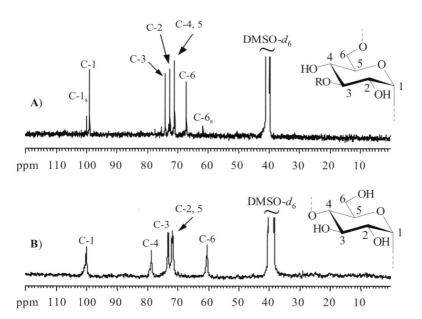

Fig. 3.3. Comparison of the ^{13}C NMR spectra of **A** dextran from *Leuconostoc* spp. (M_w 6000 g/mol, in DMSO-d_6, subscript s denotes branching points) and **B** starch (maize starch with 28 % amylose, in DMSO-d_6)

The C-1 and C-6 signals are particularly sensitive and therefore suitable to determine different substructures in polysaccharides by means of ^{13}C NMR spectroscopy. The existence of the maltotriose units in pullulan can be rapidly concluded from the occurrence both of three separate signals for the C-1 atom at 98.6, 100.6 and 101.4 ppm, and for C-6 at 60.2, 60.8 and 66.9 ppm (Fig. 3.4, assignment, see Table 3.6) [57]. In the case of well-resolved ^{13}C NMR spectra of high molecular mass dextran (M_w 60 000 g/mol), signals at 61 ppm indicate a C-6 with an unmodified hydroxyl group. A (1→3) linkage of the RU occurs, as can be concluded from a small signal for a substituted C-3 at about 77 ppm.

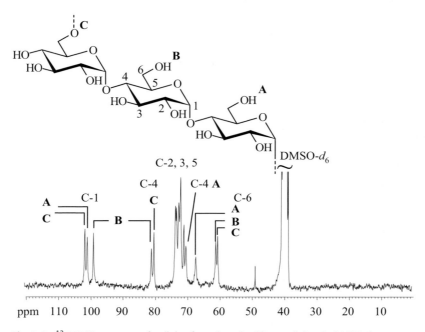

Fig. 3.4. ^{13}C NMR spectrum of pullulan from *Aureobasidium pullulans* in DMSO-d_6

Table 3.6. Assignment of ^{13}C NMR signals of pullulan in D_2O (adapted from [57])

Pullulan	Chemical shift (ppm)					
	C-1	C-2	C-3	C-4	C-5	C-6
(1→4)-(1→6)-(1→4) Glc	98.8	72.1	74.2	78.7	72.1	61.8
(1→4)-(1→4)-(1→6) Glc	100.7	72.4	74.2	78.2	72.1	61.5
(1→6)-(1→4)-(1→4) Glc	101.1	n.d.	74.0	70.6	71.3	67.6

The signal for the adjacent carbon (C-5) is shifted to lower field for carboxylic acid moieties at the polymer backbone (uronic acids), e.g. in algal polysaccharides (alginate). In alginates, the C-5 signal is found at about 78 ppm. For alginates built up of β-D-mannuronic acid and α-L-guluronic acid units, a sequence analysis can be performed. The C-5 ($\approx$ 78 ppm), C-4 ($\approx$ 81 ppm) and C-1 ($\approx$ 103 ppm) signals are strongly influenced by the sequence of the different acids (Fig. 3.5), and can be used to gain insight into the type of linkage present and the amount of the subunits. To apply ^{13}C NMR spectroscopy to alginates, it is usually necessary to employ hydrolytic degradation to obtain solutions with a reasonable viscosity. However, the degradation may lead to undesired side reactions.

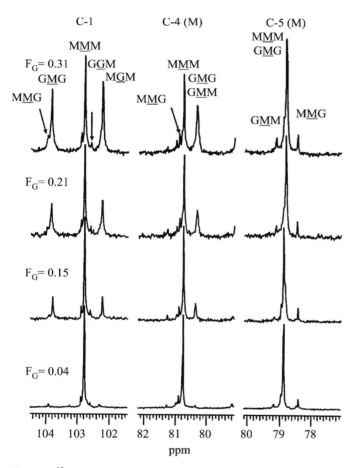

Fig. 3.5. ^{13}C NMR spectra of alginates from bacteria with different amounts of α-L-guluronic acid units. Underlined M (D-mannuronic acid) and G (L-guluronic acid) denote signals from M and G residues respectively, whereas letters not underlined denote neighbouring residues in the polymer chain. Numbers describe which proton in the hexose is causing the signal (reproduced with permission from [60], copyright The American Society for Biochemistry and Molecular Biology)

The use of solution state ^{13}C NMR spectroscopy is limited for untreated galactomannans. Nevertheless, ^{13}C NMR spectra of galactomannans from locust bean, guar and fenugreek gums have been obtained and the peaks were assigned for partially hydrolysed samples. Assignment of the splitting of the mannosyl C-4 resonances can be used for the calculation of the mannose:galactose ratios in the hydrolysed gums (Table 3.7, [56]).

Table 3.7. Assignment of ^{13}C NMR signals of galactomannan in D$_2$O (adapted from [56])

Repeating unit	Chemical shift (ppm)					
	C-1	C-2	C-3	C-4	C-5	C-6
Galactose	101.5	71.2	72.0	72.2	73.8	63.9
Mannose, unsubstituted	102.8	72.8	74.2	79.1	77.7	63.3
Mannose, substituted	102.7	72.6	74.1	79.6	76.0	69.2

For polysaccharides insoluble in DMSO or water, such as cellulose or chitin, the application of solid state ^{13}C NMR may be used. Besides the structural information on the RU and modified RU, the spectra reveal supramolecular features [61]. Owing to the supramolecular interactions, the signals are shifted generally to lower field (higher ppm), as shown in Table 3.8 for C-1, C-4 and C-6 signals of cellulose.

Table 3.8. Chemical shifts for C-1, C-4 and C-6 signals of cellulose in solid state ^{13}C NMR spectra, compared with data for cellulose dissolved in DMAc/LiCl (adapted from [62])

Polymorph	^{13}C Chemical shifts (ppm)		
	C-1	C-4	C-6
Cellulose in LiCl/DMAc	103.9	79.8	60.6
Cellulose I	105.3–106.0	89.1–89.8	65.5–66.2
Cellulose II	105.8–106.3	88.7–88.8	63.5–64.1
Cellulose III	105.3–105.6	88.1–88.3	62.5–62.7
Amorphous cellulose	ca. 105	ca. 84	ca. 63

Solid state ^{13}C NMR spectroscopy of chitin shows an upfield shift of the C-2 signal to about 58 ppm, compared to cellulose. The technique can be used to calculate the degree of *N*-acetylation from the signal ratio of the methyl moieties of the acetyl function at about 21 ppm versus the carbons of the AGU in the range 58–103 ppm [47].

For solution ^{13}C NMR investigation of cellulose and chitin, specific solvents must be applied. Cellulose can be measured in solvents, e.g. DMSO-d_6/TBAF, ionic liquids or salt melts. A spectrum of cellulose dissolved in DMSO-d_6/TBAF is shown in Fig. 2.1. The chemical shifts of the AGU carbon signals depend on the solvent used, as shown in Table 3.9 [63]. It should be noted that for solutions of polysaccharides in non-deuterated solvents, e.g. cellulose in methylimidazolium chloride, ^{13}C NMR spectroscopy needs to be carried out using a NMR coaxial tube with an interior tube filled with a deuterated liquid. ^{13}C NMR spectra of chitin can be acquired in CD_3COOD or a mixture of CD_3COOD/TFA (Table 3.10).

Table 3.9. Chemical shifts in ^{13}C NMR spectra of cellulose (DP 40), for various solvents [63, 64]

Solvent	Chemical shift (ppm)					
	C-1	C-2	C-3	C-4	C-5	C-6
NaOH/D$_2$O	104.5	74.7	76.1	79.8	76.3	61.5
Cadoxen	103.8	74.9	76.6	78.9	76.4	61.8
Triton B	104.7	74.9	76.7	80.1	76.4	61.8
LiCl/DMAc	103.9	74.9	76.6	79.8	76.6	60.6
NMMO/DMSO	102.5	73.3	75.4	79.2	74.7	60.2
TFA/DMSO	102.7	72.9	74.7	80.2	74.7	60.2
DMSO	101.8	72.6	73.7	77.3	71.7	60.2

Table 3.10. Chemical shifts of the ^{13}C NMR signals of chitin in CD_3COOD and mixture of CD_3COOD/TFA

Solvent	Chemical shift (ppm)					
	C-1	C-2	C-3	C-4	C-5	C-6
TFA/CD$_3$COOD	98.2	56.2	70.7	77.9	75.3	60.8
CD$_3$COOD	100.5	58.7	73.0	79.4	77.7	62.9

In addition to the detection of substructures, the general assignment of ^{13}C NMR spectra discussed is also useful for the elucidation of structural features of unknown polysaccharides. The assignment of the ^{13}C NMR data may be carried out by spectra simulation of an oligomer using standard software (e.g. ChemDraw Ultra Version 5.0), two-dimensional NMR techniques, and DEPT 135 NMR spectroscopy. The DEPT NMR technique reveals whether a carbon carries a proton, and the degree of protonation (CH, CH$_2$ or CH$_3$). In Fig. 3.6, a DEPT 135 NMR spectrum of dextran in DMSO-d_6 is shown as a typical example. The negative signals in the range of 60–69 ppm are caused by different CH$_2$ groups, and can therefore be assigned to differently functionalised C-6 atoms, i.e. (1→6) linkages of the main chain ($\approx$ 67 ppm), "free" C-6 atoms ($\approx$ 60 ppm) and (1→6) linkages of the substructure ($\approx$ 69 ppm, mainly functionalised C-6 adjacent to a (1→3) linkage). The assignment of ^{13}C NMR spectra via two-dimensional NMR techniques is described in Sect. 3.2.3.

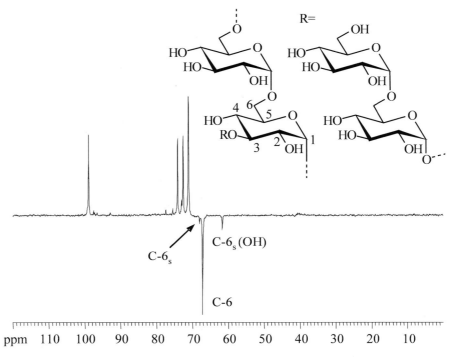

Fig. 3.6. DEPT 135 NMR spectrum of dextran showing negative signals for the different CH$_2$ groups (subscript s indicates the occurrence of the major substructure (1→3) linkage)

3.2.2 ^{1}H NMR Spectroscopy

Structural analysis using ^{1}H NMR spectroscopy is limited because of the structural diversity of the different protons along the chain, and their coupling patterns. This is displayed for a chain prolongation of cellulose oligomers and the assignment of the corresponding proton signals [65]. The complete spectroscopic data for a methyl β-D-cellohexaose is listed in Table 3.11, illustrating the enormous amount of different signals contributing to broad lines in the case of a polymer. Two separate signals are observed for H-6, due to the neighbouring chiral carbon atom of position 5.

Nevertheless, ^{1}H NMR spectroscopy is very helpful for certain tasks, e.g. determination of branching patterns and investigation of the interaction polysaccharide–solvent, because it is a fast method and the signal intensities can be used for quantification, in contrast to standard ^{13}C NMR spectroscopy. In addition, solutions with about 1% (w/w) polymer are sufficient for the acquirement of a spectrum.

Table 3.11. Signal assignment of ^{1}H NMR spectra of methyl β-D-cellohexaose (doublet, d, doublet of doublets, dd, multiplet, m, triplet, t, adapted from [65])

Ring	^{1}H (ppm)	Multiplicity	Ring	^{1}H (ppm)	Multiplicity
A			**D**		
H-1	4.40	d	H-1	4.53	d
H-2	3.30	dd	H-2	3.36	t
H-3	3.64	m	H-3	3.65	m
H-4	3.63	m	H-4	3.69	m
H-5	3.59	m	H-5	3.62	m
H-6	3.82	dd	H-6	3.83	dd
H-6′	3.99	dd	H-6′	3.98	d
B			**E**		
H-1	4.53	d	H-1	4.53	d
H-2	3.36	t	H-2	3.36	t
H-3	3.65	m	H-3	3.65	m
H-4	3.69	m	H-4	3.69	m
H-5	3.62	m	H-5	3.62	m
H-6	3.83	dd	H-6	3.83	dd
H-6′	3.98	d	H-6′	3.98	d
C			**F**		
H-1	4.53	d	H-1	4.51	d
H-2	3.36	t	H-2	3.31	dd
H-3	3.65	m	H-3	3.51	t
H-4	3.69	m	H-4	3.42	t
H-5	3.62	m	H-5	3.49	ddd
H-6	3.83	dd	H-6	3.74	dd
H-6′	3.98	d	H-6′	3.91	dd

Table 3.12 gives an overview of relevant chemical shifts in ^{1}H NMR spectra of polysaccharides and an assignment of the corresponding protons. Depending on the solvent applied, the hydroxyl groups of the polysaccharides can be detected. In pure DMSO-d_6, sharp signals in the range from about 4.4 to 5.4 are observed. They may give structural information but interfere, for a number of polysaccharide NMR spectra, with the RU proton signals. Consequently, more complex spectra result, which may complicate the structure determination of unknown polymers. In this case, the application of D$_2$O measurements in mixtures of D$_2$O/DMSO-d_6 or measurements in DMSO-d_6 after exchange of the protons against deuterons is recommended, which suppresses the OH signals.

Owing to of the splitting patterns in ^{1}H NMR spectra, illustrated for cellulose in Table 3.6, the region of the RU protons in polysaccharide spectra contains more a "fingerprint type" of information. Nevertheless, signals of the anomeric protons at position 1 are the most sensitive probe for structure elucidation. For α-D-sugar

Table 3.12. General overview of chemical shifts and the corresponding carbon atoms for ^{1}H NMR
signals of polysaccharides

H atom (moiety)	Chemical shift (ppm)
H-1 (α)	4.9–5.8
H-1 (β)	4.4–4.9
H-2 to H-6	3.3–4.5
COOH	9–13
O–CH$_3$	3.3–3.5
O–(C=O)CH$_3$	2.0–2.2
CH$_3$	1.4–1.6

moieties, signals are in the range 4.9–5.8 ppm and for β-D-sugar moieties in the
range 4.4–4.9 ppm (Table 3.15). (1→6) branching generally results in a high-field
shift in comparison to (1→4) linkage. Dextran (1→6) shows a signal for the
anomeric proton at 4.95 ppm, in contrast to starch (1→4) where it appears at
5.42 ppm. This shift is applied for the sequence analysis of pullulan. In the case of
dextran, the occurrence of α-(1→3)-D-glucosyl side chains can be concluded from
a signal at 5.28 ppm (downfield shift of about 0.3 ppm, compared to the (1→6)
main chain, [66]).

If a polymer is built up of different substructures, a quantification of structural
features, e.g. branching and chain length of side chains, can be carried out via
evaluation of the anomeric proton. This is shown by different β-D-(1→3, 1→6)
linked glucans, e.g. laminaran, curdlan and scleroglucan [55]. The relative ratios
of H-1 at different AGUs provide the information about the DP and the degree of
branching. The α- and β-anomeric protons on reducing terminals occur at 5.02–
5.03 ppm (J 3.6–3.7 Hz) and 4.42–4.43 ppm (J 7.6–7.9 Hz) respectively, whereas
the H-1 protons of internal AGUs and β-(1→6) branched AGUs appear at 4.56–
4.59 ppm (J 7.6–7.8 Hz) and 4.26–4.28 ppm (J 7.6–10.6 Hz) respectively, in a mixed
solvent of DMSO-d_6:D$_2$O (6:1) at 80 °C. In addition, the non-reducing terminal H-1
and H-1 at the AGU next to the reducing terminal can be assigned at 4.45–4.46 ppm
(J 7.8–7.9 Hz) and 4.51–4.53 ppm (J 7.8 Hz) respectively. In this way, the DP of
the laminaran was determined to be 33 and the DB 0.07. The number of glucosyl
units in the side chain of laminaran is more than one. The DB of scleroglucan was
precisely calculated as 0.33. Comparable results have been gained for the structure
analysis of mannans from yeast [57, 67]. Only α-linkages are determined from
signals in the range 5.04–5.31 ppm. The complete branching pattern and even
(1→2) linkages can be determined (see Fig. 3.7).

^{1}H NMR spectroscopy can be applied for sequence analysis of pullulan. In
addition to the evaluation of the signals of the anomeric protons, the sequence is
accessible from the other AGU proton signals and can be assigned as displayed in
Table 3.13. From the shift of the protons, the position of the corresponding AGU in

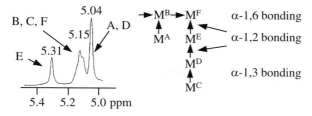

Fig. 3.7. Branching pattern of mannans from yeast and the corresponding shifts of the anomeric protons in ^{1}H NMR spectra (reproduced with permission from [51], copyright Wiley VCH and [67])

the chain can be determined. In addition to the expected subunits [(1→4), (1→6), (1→4); (1→4), (1→4), (1→6), and (1→6), (1→4), (1→4)], which represent the maltotriose unit, ^{1}H NMR spectroscopy proves that, depending on the source, pullulan may also contain α-(1→6) linked maltooligosaccharides with a (1→4), (1→4), (1→4) building unit [57].

Table 3.13. Chemical shifts of pullulan protons in D_2O (adapted from [57])

Pullulan	Chemical shift (ppm)					
	H-1	H-2	H-3	H-4	H-5	H-6
(1→4)-(1→6)-(1→4) Glc	4.94	3.59	3.99	3.65	3.84	3.82, 3.91
(1→4)-(1→4)-(1→6) Glc	5.36	3.63	3.94	3.58	3.91	3.82, 3.84
(1→6)-(1→4)-(1→4) Glc	5.32	3.60	3.69	3.45	3.91	3.97, 3.91
(1→4)-(1→4)-(1→4) Glc	5.31	3.62	3.98	3.65	3.91	n.d.

Spectra of polysaccharides containing uronic acid moieties exhibit signals for protons at position 5 in the same region as the anomeric protons (4.4–5.2 ppm). This is shown in Fig. 3.8 for the ^{1}H NMR spectrum of a degraded alginate.

Signals of the anomeric protons and the protons in position 5 of high-resolution proton NMR of slightly hydrolysed samples can be used to provide information on the L-guluronic to D-mannuronic acid ratios, to determine the frequencies for the four diads (MM, MG, GM, GG) and all triads centred on L-guluronic acid, by using the signals given in Table 3.14 [68,69].

Calculation of monad, diad and triad fractions is possible based on the following relations: $F_G + F_M = 1$, $F_G = F_{GG} + F_{GM}$, $F_M = F_{MM} + F_{MG}$, $F_{GG} = F_{GGM} + F_{GGG}$, $F_{MM} = F_{MMG} + F_{MMM}$, $F_{MG} = F_{MGM} + F_{MGG}$ and $F_{GM} = F_{GMM} + F_{GMG}$ (F values are the mole fractions, [60]).

^{1}H NMR spectra (Fig. 3.9) of depolymerised chitin and chitosan may be acquired in acidic aqueous solution at pD 3 and a temperature of 90 °C. The pD value is analogous to the pH value but corresponds to the concentration of deuterons.

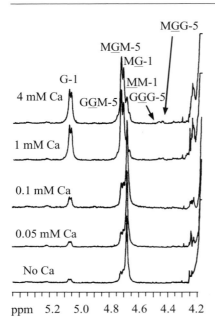

Fig. 3.8. Sequence distribution pattern of an alginate monitored by ^{1}H NMR spectroscopy. *Underlined* M (D-mannuronic acid) and G (L-guluronic acid) denote signals from M and G residues respectively, whereas *letters not underlined* denote neighbouring residues in the polymer chain. *Numbers* describe which proton in the hexose is causing the signal (reproduced with permission from [60], copyright The American Society for Biochemistry and Molecular Biology)

Table 3.14. Chemical shifts of alginate protons acquired in D$_2$O at 90 °C (reprinted from Carbohydr Res 118, Grasdalen et al., High-field, ^{1}H-n.m.r.-spectroscopy of alginate: sequential structure and linkage conformations, pp 255–260, copyright (1983) with permission from Elsevier)

Residue	Sequence	Proton			
		H-1 (reducing end)		H-1	H-5
		(α)	(β)	(internal)	(internal)
M	MM	5.21	4.89	4.67	
	MG			4.70	
G	GG		4.88	5.05	
	GM		4.84		
	GGG			5.05	4.46
	MGG			5.05	4.44
	GGM			5.05	4.75
	MGM			5.05	4.73

Depolymerisation can be achieved by treatment with nitrous acid or lysozyme, and yields information about the sequence because no significant change of the DDA appears [70]. The spectra show characteristic signals for the CH$_3$ moiety of the N-acetyl function at 2.04 ppm, and for H-2 adjacent to a deacetylated NH$_2$ moiety at 3.15 ppm. The H-1 protons at N-acetylated RU give peaks at 4.55–4.65 ppm, and at 4.85 ppm in a deacetylated aminodeoxy glucose unit. This spectral information

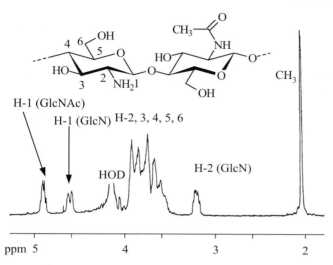

Fig. 3.9. ^{1}H NMR spectrum of chitosan in D$_2$O (pD 3 at 90 °C, reprinted from Carbohydr Res 211, Varum et al., Determination of the degree of *N*-acetylation and the distribution of *N*-acetyl groups in partially *N*-deacetylated chitins (chitosans) by high-field n.m.r. spectroscopy, pp 17–23, copyright (1991) with permission from Elsevier)

can be exploited for the determination of the *N*-acetyl content. Moreover, the spectra give insights into the linkage of acetylated (**A**) and deacetylated (**D**) RU, and may be used for sequence analysis (Fig. 3.10).

An interesting approach for the analysis of complex polysaccharide structures or mixtures of polysaccharides is the complete degradation (acid hydrolysis) and determination of the type of sugar and the concentration using the α- and β-anomeric protons (H-1) as "probes". A list of chemical shifts for these protons is given in Table 3.15. The method is fast and reliable but gives no information about the linkage of the sugar components. Nevertheless, it can successfully be applied for the analysis of lignocellulosic biomass (wood, pulp, agricultural residues, etc.) containing glucose, mannose, galactose, xylose, rhamnose, arabinose and glucuronic acid [71]. The different hydrolytic stability of the basic sugar units (see below) needs to be considered if this type of analysis is exploited.

3.2.3 Two-dimensional NMR Techniques

The complete assignment of NMR data and the verification is usually carried out by means of spectra simulation, DEPT (distortionless enhancement by polarisation transfer) NMR, and two-dimensional NMR techniques such as COSY (correlated spectroscopy, ^{1}H ^{1}H coupling), HMQC (heteronuclear multiple quantum coherence, ^{1}H ^{13}C coupling) and HMBC (heteronuclear multiple bond correlation, ^{1}H ^{13}C coupling). This is briefly described for dextran from *Leuconostoc* spp. below as an example, but can be carried out in a comparable manner for most naturally occurring polysaccharides. In the ^{13}C NMR spectra, the glycosidic carbon C-1 yields

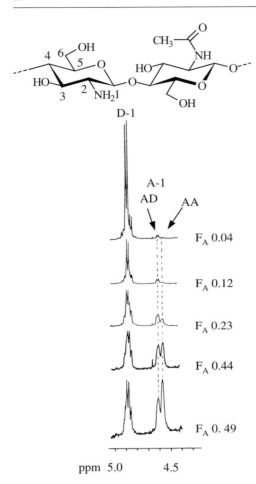

Fig. 3.10. ^{1}H NMR region of the anomeric proton of chitosan (D$_2$O, pD 3 at 90 °C) with different mole fractions of acetylated units (FA), revealing signals for the diads AD (GlcNAc-GlcN) and AA (GlcNAc-GlcNAc) usable for sequence analysis (reprinted from Carbohydr Res 211, Varum et al., Determination of the degree of N-acetylation and the distribution of N-acetyl groups in partially N-deacetylated chitins (chitosans) by high-field n.m.r spectroscopy, pp 17–23, copyright (1991) with permission from Elsevier)

a signal at about 96–103 ppm, which can be unambiguously assigned. The primary carbon (C-6) signal is determinable by DEPT 135 NMR spectroscopy because it is a CH$_2$ moiety and gives a negative peak. Using HMQC experiments, which yield information about the carbon–proton coupling over 1 bond (direct neighbourhood), the corresponding H-1 at 4.7 ppm and two signals for the chemically not equivalent protons at C-6 (H-6 and H-6′ at 3.55 and 3.7 ppm) can be found via their cross peaks (Fig. 3.11). Moreover, if DMSO-d_6 is applied as solvent, the OH moieties (at 4.1–4.65 ppm) can be identified from the HMQC because they do not give cross peaks in the spectrum.

Complete assignment of the ^{1}H NMR spectra is then possible by COSY NMR, as shown in Fig. 3.12, which contains information about the coupling of adjacent protons (three bonds apart). Starting from the signal of H-1, one can move from

Table 3.15. Chemical shifts for anomeric protons of a number of sugars used as probes for the analysis of complex polysaccharides (adapted from [19] and [72])

Monosaccharide[a]	Protons								
	H-1	H-2	H-3	H-4	H-5	H-6	H-7	CH₃	NAc
α-D-Glc-(1→	5.1	3.56	3.72	3.42	3.77	3.77	3.87	–	–
β-D-Glc-(1→	4.4	3.31	3.51	3.41	3.45	3.74	3.92	–	–
α-D-Man-(1→	1.9	3.98	3.83	3.70	3.70	3.78	3.89	–	–
βD-Man-(1→	4.7	4.04	3.63	3–58	3.37	3.76	3.93	–	–
α-D-Gal-(1→	5.2	3.84	3.90	4.02	4.34	3.69	3.71	–	–
β-D-Gal-(1→	4.5	3.52	3.67	3.92	3.71	3.78	3.75	–	–
β-D-GlcNAc-(1→	4.7	3.75	3.56	3.48	3.45	3.90	3.67	–	2.04
α-D-GalNAc-(1→	5.2	4.24	3.92	4.00	4.07	3.79	3.68	–	2.04
β-D-GalNAc-(1→	4.7	3.96	3.87	3.92	3.65	3.80	3.75	1.23	2.01
α-L-Fuc-(1→	5.1	3.69	3.90	3.79	4.1–4.9[b]	–	–	1.28	–
α-L-Rha-(1→	4.9	4.06	3.80	3.46	3.74	–	–	–	–
β-D-Xyl-(1→	4.5	3.27	3.43	3.61	[c]	–	–	1.32	–
3-O-Me-α-L-Fuc-(1→	4.8	3.70	3.40	–	3.89	–	–	1.32	–
3-O-Me-α-L-Rha-(1→	5.0	4.24	3.59	3.52	3.77	–	–	1.32	–
2,3-di-O-Me-α-L-Rha-(1→	5.1	3.94	3.52	3.41	3.73	–	–	–	–
3,6-di-O-Me-β-D-Glc-(1→	4.7	3.34	3.31	3.51	3.51	3.66	3.78	–	–

[a] Area averaged values for non-reducing terminal sugars linked by glycosidic linkage to the adjacent monosaccharide. Signals for protons at the ring carbons are shifted downfield when linked by another monosaccharide at the hydroxyl group of that carbon

[b] Signals vary considerably more than other signals, due to conformational features

[c] $H_5 ax$ 3.29; $H_5 eq$ 3.93

cross peak to cross peak (see arrows in Fig. 3.12) and can thereby determine the signals of H-2, H-3, H-4 and H-5.

Coupling between H-5 and H-6 is usually very weak and can not be found in the COSY spectra but the H-6 was already determined. Confirmation of the assignment is possible by examining the coupling of the OH protons with the AGU protons, if solutions in DMSO-d_6 are examined. In the case of dextran, only H-2, H-3 and H-4 have corresponding cross peaks. Returning to the HMQC spectrum (Fig. 3.12), the signals of the AGU carbons in the ^{13}C NMR spectrum can be unambiguously assigned.

COSY NMR spectroscopy can also be used to provide block structure information on bacterial alginates as well as information on the level of acetylation and the position of the acetyl groups within the polymer [73]. In addition to ^{1}H and ^{13}C NMR spectroscopy, ^{23}Na and ^{31}P NMR spectroscopy may be exploited for analysis of polysaccharide substructures. ^{23}Na NMR spectroscopy provides information on the rate of exchange of Na$^+$ for polymer-bound sodium ions in alginates, using

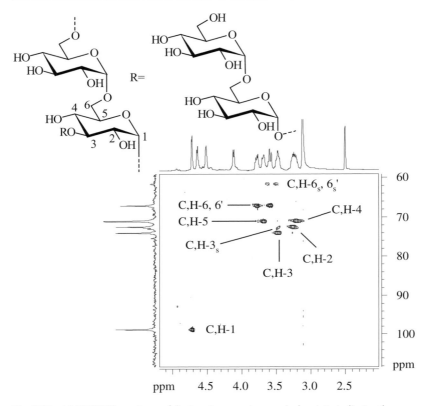

Fig. 3.11. HMQC NMR spectrum of dextran *Leuconostoc* spp. (subscript s indicates the occurrence of the major substructure (1→3) linkage)

the relaxation times of the ^{23}Na nucleus [74]. By means of ^{31}P NMR spectroscopy, phosphorylated RU in pullulan and starch have been determined [57].

3.2.4 Chromatography and Mass Spectrometry

The controlled depolymerisation and the analysis of the sugars formed by means of chromatography, preferably GLC and HPLC, is useful for structure characterisation. A simple procedure is the degradation of the polysaccharide with sulphuric acid or perchloric acid (70% w/w, see experimental section of this book) and separation of the sugars by means of HPLC (cationic exchange resin, e.g. BioRad Aminex HPX or Rezex ROA columns) using dilute sulphuric acid as eluent at elevated temperatures (65–80 °C). A summary of the retention time of different sugars commonly found in polysaccharides is given in Table 3.16. Usually, the application of a RI detector is sufficient. The response factors of the sugars should be known or need to be determined. Alternatively, pulsed amperometric detection or

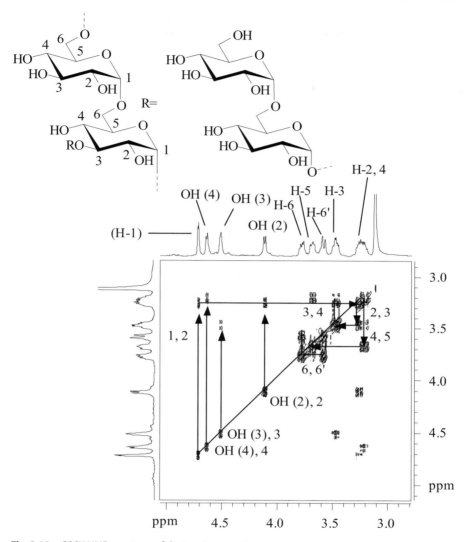

Fig. 3.12. COSY NMR spectrum of dextran *Leuconostoc* spp.

UV detection can be used. If UV detection is applied, derivatisation of the sugars is recommended (see colorimetric reactions summarised in Table 3.1).

The mutarotation does not yield a splitting of the signals but, for some ketoses such as fructose in inulin, degradation is observed if treated in strongly acidic media and gives signals that can not be assigned. A general summary of hydrolysis conditions, which can be applied for the controlled depolymerisation of polysaccharides consisting of RU with significantly different hydrolytic stability, is given in Table 3.17.

Table 3.16. Summary of retention times found by HPLC for different sugars and uronic acids commonly found in polysaccharides, with a combination of a BioRad Aminex HPX and a Rezex ROA column using 0.005 M sulphuric acid as eluent at 65 °C

Sugar	Retention time (min)
Arabinose	25.2–25.3
Fructose	23.5
Glucose	21.8–22.0
Glucuronic acid	19.5–19.7
Mannose	23.1–23.3
Rhamnose	24.6–24.7
Ribose	26.0–26.2
Xylose	23.3–23.5
(Sodium acetate)	34.8–34.9

Table 3.17. Summary of hydrolysis conditions for a variety of polysaccharides consisting of monosaccharides with different hydrolytic stability

Method	Suitable for hydrolysis of
2% oxalic acid, 80 °C, 30 min. Oxalic acid is removed by precipitation as calcium oxalate after addition of calcium acetate	Furanose is hydrolytic unstable
2 M TFA, 120 °C, 1 h. Deacetylation of GlcNAc and GalNAc results in incomplete hydrolysis, and therefore the hydrolysis can be repeated after reacetylation	Appropriate for polysaccharides with high content of Ara$p$, Xyl$p$, Man$p$, Gal$p$, Glc$p$, GlcNAc, Gal$p$NAc. Does not work for chitin and cellulose
2 M TFA, 120 °C, 1 h (incomplete hydrolysis)	GalAp, GlcAp, GalNp, GlcN are hydrolytic stable
70–90% formic acid, 100 °C, 4–24 h, followed by hydrolysis with 2 M TFA, 100 °C, 1 h (incomplete hydrolysis)	Polysaccharides with uronic acid content. Uronic acid is hydrolytic stable
72% H$_2$SO$_4$, 25 °C, 1 h, to swell the microfibres, followed by a 3–4% dilution, 120 °C, 1 h. H$_2$SO$_4$ is removed by precipitation as BaSO$_4$ after addition of Ba(OH)$_2$ and BaCO$_3$	Appropriate for crystalline polysaccharides (except cellulose)

The hydrolysis in combination with HPLC is well suited for the determination and quantification of structural features and impurities, e.g. the presence of hemicelluloses in spruce sulphite pulp, as shown in Fig. 3.13. In the case of inulin, if properly degraded with TFA, signals for the glucose and the fructose RU can be found and are used to calculate the DP of the inulin.

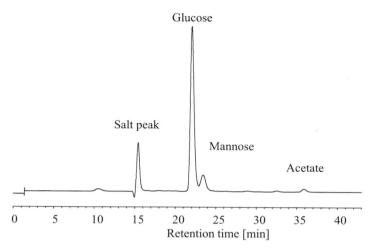

Fig. 3.13. HPLC-elugram of hydrolysed spruce sulphite pulp, showing the presence of mannose from hemicelluloses

In order to determine the ratio of the two uronic acid types in alginates, hydrolysis of the chain, followed by analysis of the sugar components are carried out. The method has practical limitations because, under acidic conditions, the RU may be subject to degradation via decarboxylation and the rates of degradation for the two uronic acids may not be equal; guluronic acid degrades faster than mannuronic acid [75]. The rates of hydrolysis also depend on the distribution of the RU along the chain [76]. Nevertheless, HPLC is very popular in the analysis of uronic acid ratios in alginate, largely due to its convenience and rapidity of use [77].

An alternative method involves the methanolysis of alginic acid, followed by analysis on an HPLC column with methanolysed D-mannurono-6,3-lactone as a standard [78]. However, the method appears to overestimate the mannuronic acid content because, due to incomplete methanolysis, the mannuronic acid residues react preferentially to the guluronic acid residues [79].

An alternative acidic treatment of the biopolymers is enzymatic digestion combined with chromatographic investigation of the resulting mixtures of oligosaccharides. Thus, the enzymatic digestion of pullulan followed by HPLC analysis of the resulting digest can be applied to determine the presence of maltooligosaccharides with more than ten glucose units [57]. For dextran, it is used for the determination of the branching pattern [66]. For the separation of oligosaccharides RP_8/RP_{18} (reversed phase silica), APS (amino propyl silica) or HPX (polystyrol with Ag^+, Pb^+, H^+ ions) phases can be exploited. Separation is also achieved using anion exchange with NaOH as eluent and a Dionex column.

For GLC studies, the sugars need to be converted to volatile derivatives. To avoid multiple peaks for the monosaccharides, modification of the C-1 aldehyde moiety is carried out prior to the derivatisation of the sugar −OH, −NH₂ or −COOH groups.

This is most frequently achieved by conversion of the aldehyde to an alditol with NaBH$_4$ in ammonia or in DMSO and alternatively by the formation of an oxime with hydroxylamine in Py or formation of a methyloxime [80].

The subsequent functionalisation of the sugars is usually acetylation with acetic anhydride in Py with sodium acetate catalysis, or silylation. For silylation, a number of reagents are useful, including hexamethyldisilazane, trimethylsilyl chloride, bistrimethylsilyl acetamide, trimethylsilyl imidazole, and bistrimethylsilyltrifluoro acetamide. Trimethylsilyl derivatives are volatile and can even be utilised for the analysis of oligosaccharides with GLC. Complications may occur in the case

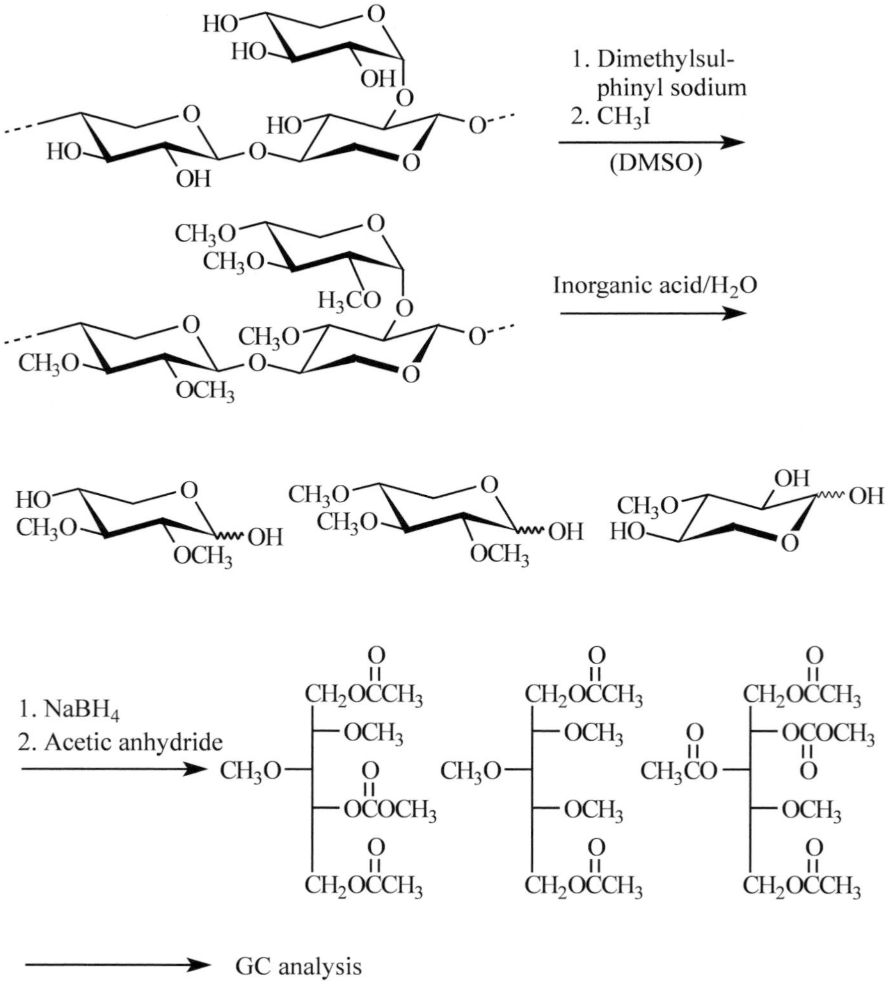

Fig. 3.14. Methylation analysis of a polysaccharide (xylan, adapted from [83])

of polysaccharides with uronic acid units. These groups are commonly converted to aldonic acids, which can be transformed to aldono-1,4-lactones. An example of the usefulness of this approach is the evaluation of the ratio of mannose:galactose in hydrolysed gums [56].

The type of linkage of RU in polysaccharide can be determined by methylation analysis in combination with GLC-MS. For this purpose, the polysaccharides are dissolved in DMSO. The methylation is carried out with Na dimsyl (prepared via partial deprotonation of DMSO with NaH or Na) and methyl iodide [81, 82]. After acidic hydrolysis of the permethylated polysaccharides, reduction of the monosaccharide derivatives formed with $NaBH_4$ and acetylation of the alditol derivatives with acetic anhydride, the GLC-MS yields information about the branching of the polysaccharide from the methoxy/acetoxy pattern. This is shown for a glucan analysis in Fig. 3.14 [83].

An alternative modification prior to chromatography yielding information about the RU of polysaccharides is the oxidation with $NaIO_4$, which selectively oxidises the carbons of vicinal OH groups to carbonyl moieties with simultaneous splitting of the C-C bond. Chromatography can be applied after reduction of the carbonyl groups with $NaBH_4$ and complete degradation [80].

Matrix assisted laser desorption/ionisation time of flight mass spectroscopy is of increasing relevance for the precise analysis of complex polysaccharide molecules, and it is a very helpful tool to analyse and quantify substructures in hemicelluloses. In the case of O-acetylated glucuronoxylans and glucomannans, the DS of acetylation can be determined. The method can also be employed to evaluate the distribution of 4-O-methyl glucuronic acid residues along the polymer chain of hardwood and softwood xylans [84].

4 Esters of Carboxylic Acids – Conventional Methods

Esterification of polysaccharides with carboxylic acids and carboxylic acid derivatives is among the most versatile transformations of these biopolymers. It gives ready synthetic access to a wide range of valuable products. Commercial processes are carried out exclusively under heterogeneous conditions, due to the high viscosity of polysaccharide solutions, the high costs of solvents, and the ease of workup procedure in the case of multiphase conversions. One aims for completely functionalised derivatives because partial conversion leads mainly to insoluble polymers, specifically in the case of cellulose.

A variety of solvents have been studied and even special solvent mixtures established for homogeneous acylation at the laboratory scale. These homogeneous reactions permit synthesis of highly soluble, partially derivatised polymers and are the prerequisite for the application of "state of the art" organic reagents yielding broad structural diversity. For cellulose and chitin, the development of novel solvents heralded a new era of bio-based functional polymers. Both common organic solvents and multicomponent solvents are still widely studied for esterification procedures yielding novel structures.

4.1 Acylation with Carboxylic Acid Chlorides and Anhydrides

Conventional esterifications of polysaccharides are acylation procedures developed as heterogeneous processes, but now include homogeneous mixtures during the esterification, caused by the dissolution of the esterified product, applying usually carboxylic acid anhydrides or chlorides. In the case of sensitive acids, these reactive compounds are either expensive or inaccessible, and the anhydrides and chlorides of more complex acids are insoluble. Thus, conventional acylation is applied for the complete conversion of all hydroxyl groups of the polysaccharide with aliphatic (acetate to stearate) and aromatic acids (substituted benzoic acids). Over the last 60 years, an enormous amount of papers have been published (approximately 54 000 papers dealing with cellulose esters alone), and this chapter presents an overview of present general techniques and their specific potential.

4.1.1 Heterogeneous Acylation – Industrial Processes

The most common method for the acylation of polysaccharides is the reaction with carboxylic acid anhydride in heterogeneous phase (Fig. 4.1).

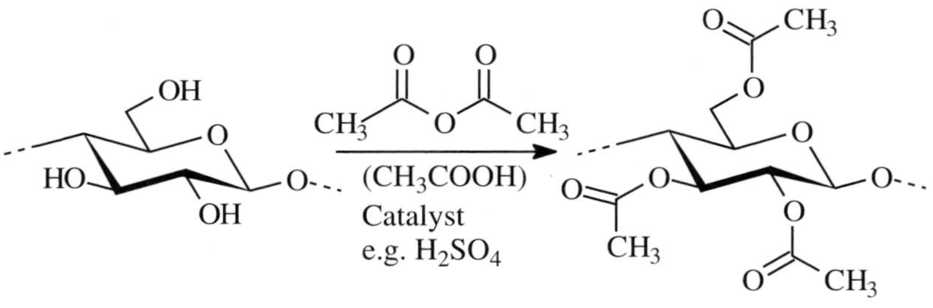

Fig. 4.1. Scheme of the conversion of cellulose with acetic anhydride/acetic acid

Cellulose acetate is the most commercially important polysaccharide ester of a carboxylic acid, and is prepared industrially or at the laboratory scale by conversion of cellulose with a mixture of acetic acid and acetic anhydride (10–40% excess to the amount needed for cellulose triacetate formation) in the presence of sulphuric acid as catalyst (up to 15%, w/w). It is assumed that intermediately the mixed sulphuric acid–acetic acid anhydride is formed (usually called acetyl sulphuric acid) generating an acetyl cation (Fig. 4.2, [85, 86]). Hence, during this conversion, a partial sulphation is observed that suppresses the formation of a real cellulose triacetate. Nevertheless, most of the sulphate groups introduced are exchanged by acetyl functions during the reaction or split off during the workup. In the case of starch, a dehydration (water content less than 3%) and conversion at reflux temperature is recommended for the formation of a starch triacetate [87].

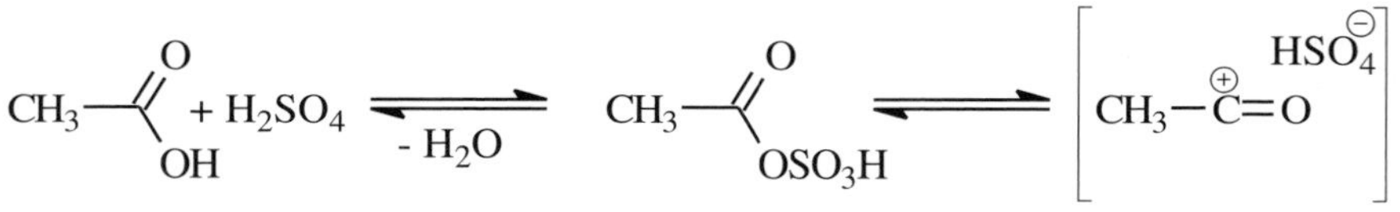

Fig. 4.2. Formation of the reactive, intermediately formed acetyl sulphuric acid

For better control of the reaction temperature and to diminish the amount of catalyst (as low as 1%, w/w H_2SO_4), acetylation can be carried out in methylene chloride, which is combined with the dissolution of the products formed in the final phase of the reaction. Most commercial cellulose acetate is produced via this route.

An alternative is the acetylation on the intact cellulose fibre (fibre acetylation) in an inert solvent such as toluene with perchloric acid as highly efficient catalyst [88]. The triester is obtained applying only 15 mg catalyst on 5 g cellulose in 40 ml acetylation mixture within 24 h at 32 °C. Moreover, the introduction of additional ester groups is avoided, as observed for sulphuric acid catalysis. This method is exploited if the superstructure of the polysaccharide should be retained, which is essential for a number of applications, e.g. as solid phase for chiral separation in chromatographic methods (see Chap. 10).

Fully acetylated cellulose can be partially deacetylated in a one-pot hydrolysis to give the widely applied acetone soluble cellulose diacetate (acetyl content ca. 40%; DS_{Ac} 2.4–2.6) indispensable for spinning or shaping processes. This "synthetic detour" is necessary because cellulose acetate samples with a comparable DS synthesised directly from cellulose are not soluble in acetone [1, 2]. Reasons for this behaviour are still not completely understood (see Chap. 8). Both an uneven distribution of the ester functions along the chain or the complete functionalisation of position 6 may contribute to the insolubility of cellulose diacetate in acetone. A comparable behaviour is observed for the water solubility of cellulose acetate with DS values between 0.6–0.9, which can be achieved only by homogeneous esterification (Table 4.1) or hydrolysis of cellulose triacetates [89].

The solubility over the whole range of DS values is given in Table 4.1. The synthesis of polysaccharide esters of longer aliphatic acids up to butyrate and the mixed esters with acetates are basically accessible via the same path, i.e. treatment of the polymer in methylene chloride with acetic acid and of the corresponding anhydrides with sulphuric acid as catalyst.

Table 4.1. Solubility (– insoluble, + soluble) of cellulose acetate (obtained by hydrolysis of cellulose triacetate) depending on the DS values

Cellulose acetate DS	Solvent Chloroform	Acetone	2-Methoxy-ethanol	Water
2.8–3.0	+	–	–	–
2.2–2.7	–	+	–	–
1.2–1.8	–	–	+	–
0.6–0.9	–	–	–	+
< 0.6	–	–	–	–

A variety of alternative catalysts are available with higher efficiency, which permit milder reaction conditions. The "impeller" method is helpful in polysaccharide acylation. The carboxylic acids or their anhydrides are converted in situ to reactive mixtures of symmetric and mixed anhydrides (Fig. 4.3). Chloroacetyl-, methoxyacetyl- and, most important, trifluoroacetyl moieties are used as impellers [90, 91]. Carboxylic acid esters of polysaccharides with almost complete functionalisation can be obtained. Thus, chloroform-soluble dextran stearates and dextran myristates with DS 2.9 are prepared by treating dextran in chloroacetic anhydride with the corresponding acids at 70 °C for 1 h. The presence of magnesium perchlorate as catalyst is necessary [92].

The reactions succeed with diminished chain degradation if TFAA is used as impeller reagent. An introduction of impeller ester functions in the polysaccharide, i.e. trifluoroacetyl moieties are not found. A rather dramatic decrease in reactivity is observed for the introduction of esters in the case of cellulose in the order acetic > propionic > butyric acid [93, 94].

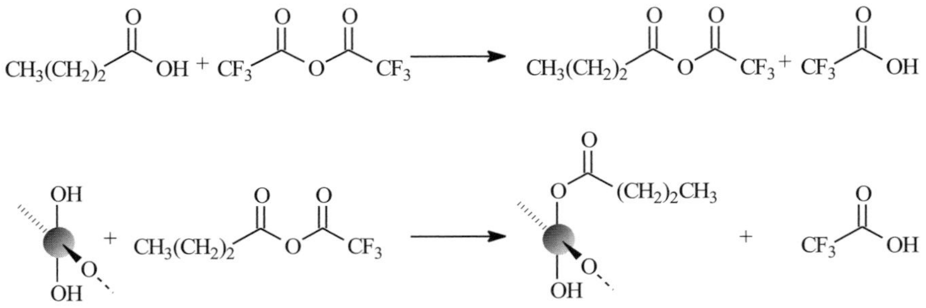

Fig. 4.3. Acylation of polysaccharides via reactive mixed anhydrides (impeller method)

Highly functionalised long-chain aliphatic acid esters of cellulose are accessible by simply mixing carboxylic acid with TFAA for 20 min at 50 °C and treating the dried cellulose at 50 °C for 5 h [95]. A summary of the DS values achieved and M_w are given in Table 4.2.

Table 4.2. DS and M_w values for long-chain aliphatic acid esters of cellulose obtained via the impeller method applying TFAA (adapted from [95])

Acid moiety	Number of carbons	DS	M_w (10^5 g/mol)
Acetate	2	2.8	–
Propionate	3	3.0	1.48
Butyrate	4	2.8	1.77
Valerate	5	2.8	2.15
Hexanoate	6	2.8	2.15
Enanthate	7	3.0	2.07
Octanoate	8	2.8	2.03
Pelargonate	9	2.9	3.54
Decanoate	10	2.9	2.32
Laurate	12	2.9	2.18
Myristate	14	2.9	2.87
Palmitate	16	2.9	3.98
Stearate	18	2.9	6.91

The impeller method can be applied for the synthesis of starch acetates and cellulose benzoates with complete functionalisation. The reaction is completed if the polysaccharide dissolves in the reaction mixture (after about 75 min at 60 °C, [96]).

An interesting new catalyst usable for polysaccharide modification with anhydrides is *N*-bromosuccinimide. The inexpensive and commercially available reagent shows high efficiency for the catalysis of the acetylation of hemicelluloses

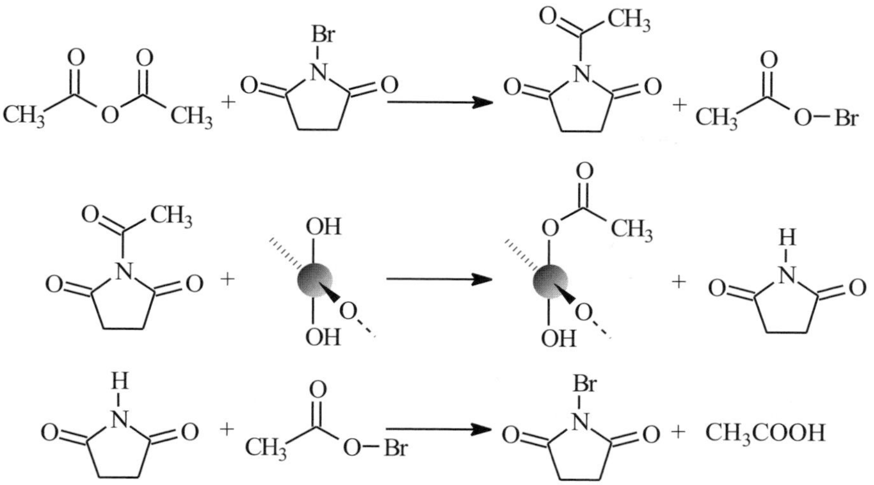

Fig. 4.4. Proposed mechanism for the acetylation of hemicelluloses with *N*-bromosuccinimide as catalyst (adapted from [98])

with the corresponding anhydride [97, 98]. A DS of 1.15 is obtained after 2 h at 80 °C, with 1% NBS as catalyst. Besides the mechanism proposed (Fig. 4.4), the activating role of intermediately formed HBr is discussed.

Titanium-(IV)-alkoxide compounds, such as titanium-(IV)-isopropoxide, are useful as esterification catalysts for the conversion of long-chain fatty acids [99]. It has been shown that the catalyst is efficient for the preparation of partially esterified cellulose derivatives, if an appropriate solvent is used (Table 4.3).

Table 4.3. Long-chain mixed cellulose esters synthesised by titanium-(IV)-isopropoxide-catalysed reaction in DMAc (adapted from [99])

| Reaction conditions | | | | Product | |
Carboxylic acid anhydride	Equivalent per AGU	Time (h)	Temp. (°C)	DS	$M_w\ 10^3$ (g/mol)
Acetic/hexanoic	2.0/2.0	9	155	1.91/0.75	164
Acetic/nonanoic	2.0/2.0	11	145	2.03/0.70	177
Acetic/lauric	3.5/1.0	12	140	2.40/0.20	295
Acetic/palmitic	2.0/2.0	12	145	2.06/0.42	125
Acetic/nonanoic	3.0/1.0	8	145	2.44/0.26	220

4.1.2 Heterogeneous Conversion in the Presence of a Base

Esterification reactions with carboxylic acid anhydrides under acidic catalysis are combined with chain degradation. This side reaction is used to adjust the DP of the products. Commercial cellulose acetates have DP values in the range 100 to 360. If the degradation is to be suppressed, esterification with the anhydride in a tertiary base, commonly Py or TEA, is recommended. The base represents the slurry medium (combined with swelling of the polysaccharide) and the acylation catalyst. The triacetate of cellulose is obtained after comparably long reaction times of 6 to 10 days at 60 °C. The same procedure with starch leads to starch triacetates; the reaction time necessary for the formation of the starch triacetate can be shortened to 24 h by increasing the temperature to 100 °C [100]. Another tool to increase the reactivity of this system is the addition of DMAP, which increases the rate of the reaction by up to 10^4 times. The catalytic efficiency is probably due to the stabilisation of the acylpyridinium ion, which plays an important role in the catalytic cycle (Fig. 4.5). Steric effects, the donor ability of the amine substituent, and the good nucleophilic properties of DMAP additionally affect the reactivity [101].

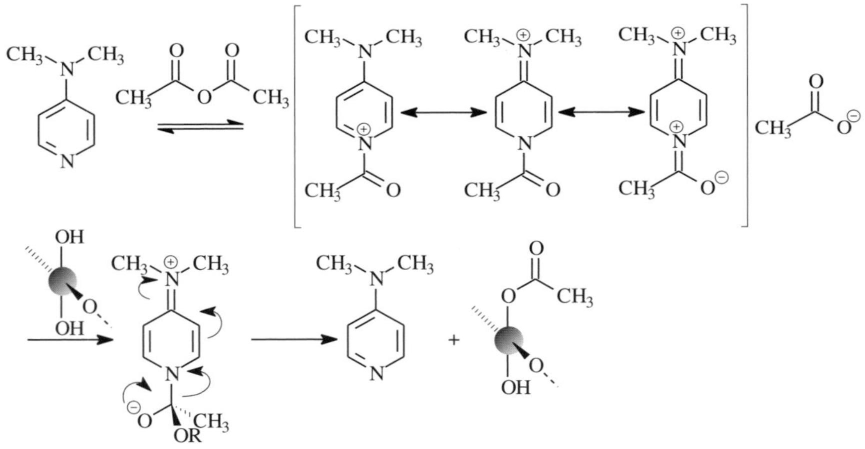

Fig. 4.5. Mechanism of the DMAP catalysis, R = polysaccharide backbone (adapted from [101])

The preparation of pullulan nonaacetate (all OH groups of the monomeric unit, see Chap. 2, are esterified) is possible within 2 h at 100 °C using this type of conversion [102]. In the case of chitin (DDA 0.16), complete acetylation both of the OH– and the NH_2 groups can be achieved within 48 h at 50 °C [103]. A comparable conversion of chitin in methanol as slurry medium yields only N-acetylation (Fig. 4.6).

Moreover, introduction of ester functions of dicarboxylic acids or mixed derivatives containing aliphatic ester moieties and ester functions of dicarboxylic acids, which leads to products with controlled release properties (see Chap. 10), can be obtained by conversion of the polysaccharide or the derivative, e.g. cellulose acetate with a dicarboxylic acid anhydride such as phthalic anhydride in Py (Fig. 4.7, [104]).

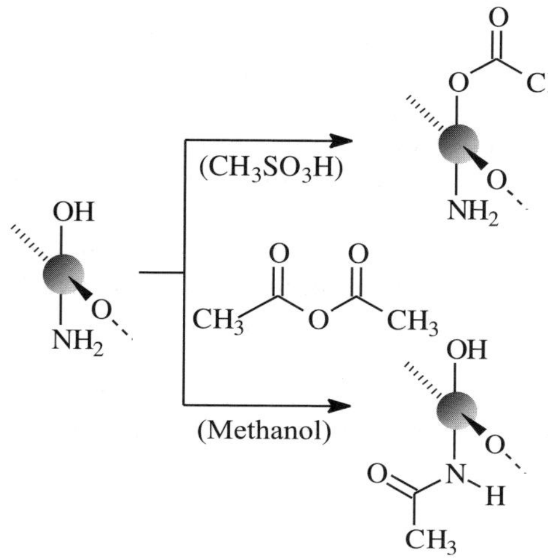

Fig. 4.6. Selective acetylation of chitin (chitosan) in different reaction media

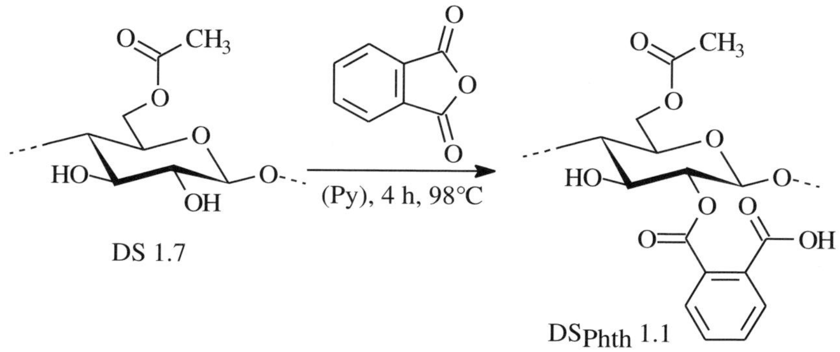

Fig. 4.7. Synthesis of a mixed cellulose acetate phthalate

Interestingly, 2-aminobenzoic acid esters of cellulose are accessible by conversion with isatoic anhydride [105]. A comparable esterification method of starch with isatoic anhydride in DMSO in the presence of TEA or in water/NaOH is possible (see Fig. 5.1). This is a readily available method to introduce an amino function into the polymer backbone.

More suitable for the acetylation of easily soluble polysaccharides (e.g. dextran, inulin, curdlan) at the laboratory scale is still the homogeneous esterification in formamide, DMF, DMSO or water, as discussed in Sect. 5.1.

For the introduction of more complex carboxylic acid moieties, i.e. fatty acid moieties, alicyclic groups or substituted aromatic functions, anhydrides are not reactive enough and insoluble in organic media. In these cases, acid chlorides in combination with a base are applied. The heterogeneous conversion of polysaccharides in a slurry consisting completely or partially (in combination with a second organic liquid) of a tertiary base, usually Py or TEA, is still a widely used practice for the preparation of fatty acid esters of polysaccharides useful as thermoplastic materials. A summary of such esters is listed in Table 4.4. The reactions lead to highly functionalised esters, which are soluble in non-polar solvents, e.g. chloroform. Nevertheless, partial functionalisation is observed for conversion at lower temperature, yielding hardly soluble products. In contrast to the synthesis of C_2 to C_4 acid esters, solubility does not present a problem because fatty acid esters of polysaccharides are prepared as derivatives that can be thermally processed. The reactive species formed in this process is an acylium salt (Fig. 4.8).

Table 4.4. Esterification of different polysaccharides with acyl chlorides in a tertiary base (Py)

Polysaccharide Reference	Carboxylic acid chloride	Conditions Molar ratio		Temp. (°C)	Time (h)	Product DS
		AGU	Reagent			
Starch [106]	Octanoyl	1	2.3	105	3	1.8
	Octanoyl	1	4.6	105	3	2.7
	Lauroyl	1	4.6	105	3	2.7
	Stearoyl	1	2.3	105	3	1.8
	Stearoyl	1	4.6	105	3	2.7
Inulin [107]	Caproyl	1	3.0	40	24	2.5
	Capryloyl	1	0.5	40	24	0.5
	Stearoyl	1	2.0	60	24	1.8
	Stearoyl	1	1.0	60	24	0.8

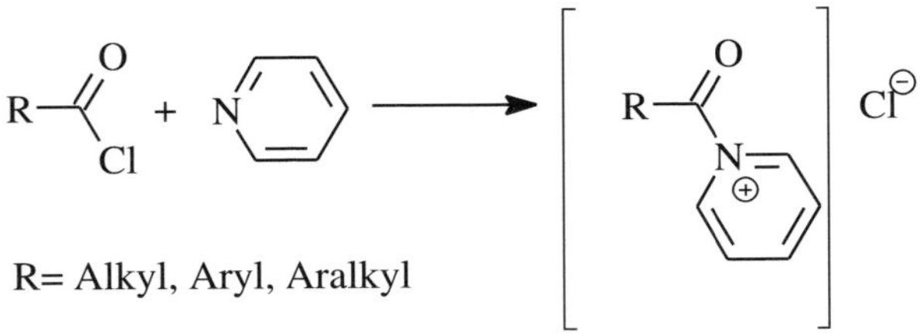

R= Alkyl, Aryl, Aralkyl

Fig. 4.8. Acylium salt formed from an acid chloride and the tertiary base Py

The Py-containing reaction mixture and the resulting crude products are mostly brown because of side reactions, mainly polycondensation. The products can be purified carefully by washing with ethanol, Soxhlet extraction with ethanol, or reprecipitation from chloroform in ethanol. The formation of 2-methyl-3-oxo-pentanoyl groups can accompany propionylation, through the mechanism shown in Fig. 4.9, leading to unreasonable DS values [108].

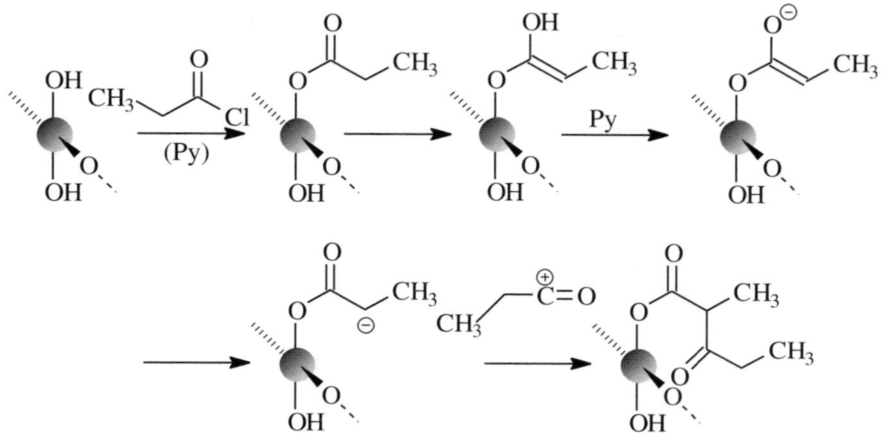

Fig. 4.9. Reaction mechanism for the formation of 2-methyl-3-oxo-pentanoyl groups during the propionylation of polysaccharides with propionyl chloride in Py (adapted from [108])

The esterification of dextran, starch and cellulose with acid chlorides is combined with less side reactions if conducted in DMF or DMAc as slurry medium or homogeneously in DMAc/LiCl without an additional base (see Sect. 5.1). Thus, pure cellulose pentanoates and hexanoates can be obtained in DMF without Py. The utilisation of an excess of base diminishes the DS values, as can be seen in Table 4.5, which summarises results for the acylation of cellulose in Py with propionyl chloride [109].

Accordingly, acylation is best performed in mixtures of a base and a diluent. Table 4.6 gives an overview of results for the cellulose propionylation with different diluents and bases, showing that the use of Py in combination with 1,4-dioxane, chlorobenzene and toluene yields efficient slurry media.

Dextran [110], cellulose [111] and starch [112] can be efficiently acylated via this route, as shown for a number of polysaccharide fatty acid esters in Table 4.7.

Highly functionalised starch palmitate and starch stearate [117], in addition to long-chain fatty acid cellulose esters from soybean fatty acids, can be obtained applying the DMF/Py mixture [118]. To increase the reactivity of these acylation systems, the application of DMAP is useful, as shown for the anhydrides (see Fig. 4.5).

Polysaccharide tribenzoates and substituted benzoic acid esters are basically accessible via the same path, i.e. conversion of the polysaccharide in Py with benzoyl chloride, but they have found no pronounced interest. Only aromatic

Table 4.5. Acylation of cellulose in Py with propionyl chloride at 100 °C for 4 h (adapted from [109])

Molar ratio			Product	
AGU	Py	Propionyl chloride	DS	Solubility in $Cl_2CHCHCl_2$
1	27.6	4.5	1.86	–
1	18.9	6.0	2.66	+
1	12.0	6.0	2.80	+
1	12.0	4.5	2.70	+
1	9.9	9.0	2.13	+
1	7.5	6.0	2.89	+
1	6.0	4.5	2.81	+
1	4.8	4.5	2.86	+
1	3.0	4.5	2.89	+
1	1.5	4.5	2.84	+

Table 4.6. Cellulose propionylation with different diluents and tertiary amine using 1.5 mol propionyl chloride/mol AGU at 100 °C

Conditions			Product
Medium	Base	Time (h)	DS
Dioxane	Py	4	2.81
Dioxane	β-Picoline	4	2.70
Dioxane	Quinoline	4	2.18
Dioxane	Dimethylaniline	48	1.57
Dioxane	γ-Picoline	24	Negligible
Chlorobenzene	Py	4	2.86
Toluene	Py	4	2.30
Tetrachloroethane	Py	24	2.23
Ethyl propionate	Py	5	2.16
Isophorone	Py	4	1.89
Ethylene formal	Py	22	0.34
Propionic acid	Py	5	0.20
Dibutyl ether	Py	22	Negligible

Table 4.7. Polysaccharide esters synthesised in slurry containing an inert organic solvent and Py, and using acid chlorides

Polysaccharide	Fatty acid moiety	Molar ratio			Organic liquid	Time (h)	Temp. (°C)	DS	Ref.
		AGU	FACl	Py					
Starch	Butyrate	1	4.5	6.2	Dioxane	6	100	> 2.8	[113]
Starch	Caproate	1	4.5	6.2	Dioxane	6	100	> 2.8	[113]
Starch	Valerate	1	4.5	6.2	Dioxane	6	100	> 2.8	[113]
Starch	Laurate	1	4.2	6.3	Dioxane	6	100	2.9	[114]
Starch	Oleate	1	3.0	3.0	Toluene	5.5	90	3.0	[115]
Starch	Myristate	1	4.5	6.0	Toluene/Dioxane	6	100	2.6	[112]
Starch	Palmitate	1	4.5	6.0	Toluene/Dioxane	6	100	2.1	[112]
Starch	Oleate	1	4.5	6.0	Toluene/Dioxane	6	100	2.7	[112]
Cellulose	Pentanoate	1	4.5	1.0	Dioxane	17	80	2.1	[116]
Cellulose	Hexanoate	1	4.5	1.0	Dioxane	17	80	2.5	[116]
Cellulose	Octanoate	1	4.5	1.0	Dioxane	17	80	2.4	[116]
Cellulose	Nonanoate	1	4.5	1.0	Dioxane	17	80	2.4	[116]
Dextran	Palmitate	1	3.0	8.0	Toluene	1.5	105	2.9	[110]

carboxylic acid esters with unsaturated side chains, e.g. cinnamic acid esters, are used for cross-linking and grafting reactions (see Chap. 10).

Schotten-Baumann reaction with acyl chlorides after activation of polysaccharides with aqueous NaOH is possible but scarcely applied today. An interesting new path is the esterification of cellulose with acyl chlorides at elevated temperatures, using vacuum to remove the HCl liberated during the reaction [118].

Moreover, heating of mixtures of dry corn starch, glacial acetic acid and carboxylic acid anhydrides under pressure in small (60 µl) stainless steel sealed pans yields starch esters with remarkable DS values. Starch acetates of DS 0.5–2.5 are obtained at temperatures of 160–180 °C within 2–10 min, with almost complete conversion. Reaction rates increase with increasing acetic acid concentration and decrease with increasing acetic anhydride amounts. The acetic acid remaining in the samples can be completely removed by vacuum stripping at 120–190 °C. Starch succinates of DS 1.0–1.5 are accessible under similar conditions. Longer reaction times (20–60 min) are required for the preparation of starch octenylsuccinates and dodecenylsuccinates having moderate DS values (≈ 0.5). Longer heating results in significant degradation [120].

The methods described above are well-established, reproducible tools for the synthesis of defined polysaccharide esters of pronounced commercial importance. Still, these synthesis methods are limited for the preparation of common aliphatic and aromatic carboxylic acid esters. To achieve acylation, a broad variety of new synthesis paths are under investigation, as described in Chap. 5.

5 New Paths for the Introduction of Organic Ester Moieties

Both the investigation of new solvents and the adaptation of esterification methodologies used in peptide synthesis have driven the new synthetic paths for carboxylic acid ester formation. The introduction of organic solvents such as DMSO, formamide and DMF, and combinations of these solvents with LiCl for dextran, pullulan and curdlan, and DMAc/LiCl and DMSO/TBAF for cellulose and starch have made the homogeneous esterification into an efficient synthesis path using dehydrating agents, e.g. DCC and CDI. The solvents and the reagents used are discussed with focus on the preparation of cellulose acetate as basic reaction but, in addition, a broad variety of specific esterification reactions is given to illustrate the enormous structural diversity accessible by these new and efficient methods.

5.1 Media for Homogeneous Reactions

Homogeneous reaction conditions are indispensable for the introduction of complex and sensitive ester moieties because they provide mild reaction conditions, selectivity, and a high efficiency. In contrast to heterogeneous processes, they can be exploited for the preparation of highly soluble, partially substituted derivatives because these conditions guarantee excellent control of the DS values. Moreover, they may lead to new patterns of substitution for known derivatives, compared to heterogeneous preparation. In addition to formamide, DMF, DMSO and water, which are good solvents for the majority of polysaccharides (Table 5.1), new solvents have been developed especially for cellulose, with its extended supramolecular structure.

A summary of cellulose solvents used for acetylation is given in Table 5.2. The dissolution process destroys the highly organised hydrogen bond system surrounding the single polysaccharide chains.

Although a wide variety of these solvents have been developed and investigated in recent years [122], only a few have shown a potential for a controlled and homogeneous functionalisation of polysaccharides. Limitations of the application of solvents result from: high toxicity; high reactivity of the solvents, leading to undesired side reactions; and the loss of solubility during reactions, yielding inhomogeneous mixtures by formation of gels and pastes that can hardly be mixed, and even by formation of de-swollen particles of low reactivity, which precipitate from the reaction medium.

Table 5.1. Solubility of polysaccharides in DMSO, DMF and water

Polysaccharide	Solubility in			
	DMF	DMSO	Py	H_2O
Cellulose	–	+ (TBAF)	–	–
Chitin	–	–	–	–
Starch	–	+ (80 °C)	–	–[a]
Amylopectin	–	+ (80 °C)	–	+
Curdlan	–	+	–	–
Schizophyllan	+ (80 °C, LiCl)	+	–	–
Scleroglucan	–	–	–	–
Pullulan	+ (80 °C)	+	+[b]	+
Xylan	+ (LiCl)	+	–	– (NaOH)
Guar	–	–	–	–
Alginate	–	–	–	+
Inulin	+	+	+	+
Dextran	+ (LiCl)	+ (40 °C)	–	+[c]

[a] Amylose is water soluble at 70 °C
[b] Depending on the source
[c] The crystalline form is insoluble [121]

Table 5.2. Solvents and reagents exploited for the homogeneous acetylation of cellulose

Solvent	Acetylating reagent	DS_{max}	Ref.
N-Ethylpyridinium chloride	Acetic anhydride	Up to 3	[123]
1-Allyl-3-methyl-imidazolium chloride	Acetic anhydride	2.7	[124]
N-Methylmorpholine-N-oxide	Vinyl acetate	0.3	[125]
DMAc/LiCl	Acetic anhydride	Up to 3	[126]
	Acetyl chloride	Up to 3	[127]
DMI/LiCl	Acetic anhydride	1.4	[128]
DMSO/TBAF	Vinyl acetate	2.7	[129]
	Acetic anhydride	1.2	[27]

5.1.1 Aqueous Media

Water dissolves or swells most of the polysaccharides described here (see Table 5.1). Thus, water can be used both as solvent for homogeneous reactions and as slurry medium. Manageable solutions are obtained for starch with a high amylopectin content, scleroglucan, pullulan, inulin and dextran by adding small amounts of the polysaccharide to water under vigorous stirring, and heating the mixture to 70–80 °C. If the viscosity of a low-concentrated solution, especially for high-molecular mass starch, guar gum and alginates, is still too high for a conversion, then an acidic (see Table 3.17 in Sect. 3.2.4) or enzymatic pre-treatment for partial chain degradation is necessary, as described for starch in Chap. 12. Despite the fact that water is commonly not an appropriate medium for esterification reactions, a number of polysaccharide esters may be obtained in this solvent. Especially starch acetates are manufactured in aqueous media by treatment with acetic anhydride. This type of conversion is used for the preparation of water-soluble starch acetate (DS 0.1–0.6), applicable in the pharmaceutical field ([130], see Chap. 10). By reacting enzymatically degraded starch (M_w 430 000 g/mol) in aqueous media with acetic anhydride in the presence of dilute (1 N) NaOH, starch acetate is obtained. The pH should be kept in the range 8.0–8.5 by stepwise addition of the base during the synthesis at RT. The preparation of completely functionalised corn starch or potato starch acetate is achieved with an excess of acetic anhydride (4-fold quantity) in the presence of 11% NaOH (w/w in the mixture added as a 50% solution). After 3 h, starch acetate with DS 2 is isolated by pouring the reaction mixture into ice water. A longer reaction time (5 h) results in complete functionalisation [131, 132]. In addition, the synthesis of starch propionates and butyrates with DS 1–2 is realised by mixing starch in water with the corresponding anhydrides and 25% NaOH for 4 h at 0–40 °C [133].

The synthesis of starch methacrylates has also been reported [134]. If untreated, native starch is used as starting material and the mixture water/starch is thermally treated, the conversion is heterogeneous, and the products are isolated by filtration. Starch 2-aminobenzoates are accessible by conversion with isatoic anhydride in the presence of NaOH (Fig. 5.1, [135]).

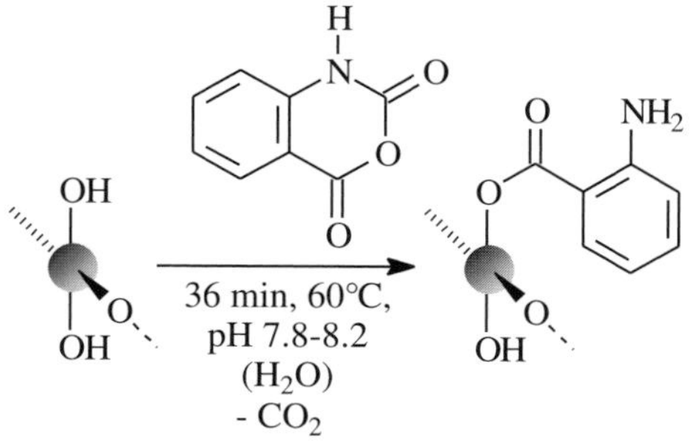

Fig. 5.1. Synthesis of starch 2-aminobenzoates in an aqueous medium

A series of starch esters with different carboxylic acid moieties (C_6–C_{10}) and moderate DS values can be prepared by acylation of the gelatinised biopolymer with the corresponding acid chloride in 2.5 M aqueous NaOH solution, which represents an economical and easy method for the starch acylation. The alkali solution acts as the medium for the derivatisation, and ensures uniform substitution. Successful esterification is limited to acid chlorides containing between 6 and 10 carbon atoms. The dependence of the DS on the chain length of the acid chloride applied is displayed in Fig. 5.2. Shorter ($< C_6$) or longer ($> C_{10}$) acid chlorides do not react under these conditions, as can be confirmed by FTIR spectroscopic and elemental analyses as well as by intrinsic viscosity analysis [136].

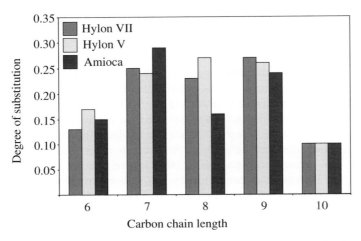

Fig. 5.2. DS values achieved by modification of starch with acid chlorides in aqueous media, in function of the chain length of the acid moieties and the starch type (amylose content: 70%, Hylon VII, 50%, Hylon V, 1%, Amioca, adapted from [136])

Acylation in aqueous media with aromatic acid chlorides, e.g. benzoyl chloride [137] or acyl imidazolides, can be carried out as well [138]. The imidazolides can be prepared in situ from the carboxylic acid with CDI or from the acid anhydride or the chloride using imidazole (see Sect. 5.2.3). The introduction of acyl functions up to stearates is achieved in water with rather low DS values, giving starch derivatives with modified swelling behaviour (Table 5.3).

Table 5.3. Starch esters prepared in water, using the carboxylic acid imidazolide (adapted from [138])

Acylating agent Imidazolide	Amount (%)[a]	pH	Time (h)	DS_{Acyl}
Acetic acid	6	8	2.0	0.06
Benzoic acid	7	8	2.0	0.05
Acrylic acid	7	8	1.8	0.01
Stearic acid	20	9	18.0	0.03

[a] Amount of acylating agent in % (w/w) in relation to starch

Aqueous media are useful for the derivatisation of hemicelluloses. For wheat straw hemicelluloses, a reaction with succinic anhydride in aqueous alkaline media for 0.5–16 h at 25–45 °C and a molar ratio of succinic anhydride to AXU of 1:1–1:5, succinoylation yields DS values ranging from 0.017 to 0.21. The pH should be in the range 8.5–9.0 during the reaction [139].

Interestingly, conversion of inulin in water using carboxylic acid anhydrides is achieved in the presence of ion exchange resins. Acetates with DS 1.5 and propionates with DS 0.8 can be isolated by filtration of the resin and vacuum evaporation of the solvent. The easy workup is limited by partial regeneration of the polymer at the resin, resulting in rather poor yields of 40–50% [107].

Molten inorganic salt hydrates have gained some attention as new solvents and media for polysaccharide modification. Molten compounds of the general formula $LiX \times H_2O$ ($X^- = I^-, NO_3^-, CH_3COO^-, ClO_4^-$) were found to dissolve polysaccharides including cellulose with DP values as high as 1500 [140–142]. Acetylation can be performed in NaSCN/KSCN/LiSCN $\times$ 2 H_2O at 130 °C, using an excess of acetic anhydride (Table 5.4). DS values up to 2.4 are accessible during short reaction times (up to 3 h). The reaction is unselective, in contrast to other esterification processes. X-ray diffraction experiments show broad signals, proving an extended disordered morphology. This structural feature imparts a high reactivity towards solid–solid reactions, e.g. blending with other polymers. Furthermore, the cellulose acetates synthesised in molten salt hydrates show low melting points, obviously because of the amorphous morphology.

Table 5.4. Experimental data and analytical results for the acetylation of cellulose in NaSCN/KSCN/LiSCN $\times$ 2 H_2O with acetic anhydride

Reaction conditions Molar ratio		Time (h)	Partial DS at		Σ
AGU	Acetic anhydride		O-6	O-2 and 3	
1	100	3.0	0.91	1.57	2.41
1	100	1.0	0.86	1.12	1.98
1	100	0.5	0.39	0.85	1.23
1	75	0.5	0.51	0.50	1.02

In common aqueous polysaccharide solvents, i.e. Cuen or Nitren, hydrolysis of the agents is competing against esterification, leading to low yields.

5.1.2 Non-aqueous Solvents

DMSO can be conveniently handled because it is non-toxic (LD_{50} (rat oral) = 14 500 mg/kg) and has a high boiling point (189 °C). During simple esterification reactions, e.g. with anhydrides, it is chemically inert. For more complex reactions,

DMSO can act as an oxidising reagent and shows decomposition to a variety of sulphur compounds. This is illustrated in Fig. 5.3 for a general DMSO-mediated oxidation of an alcohol and for the Swern oxidation.

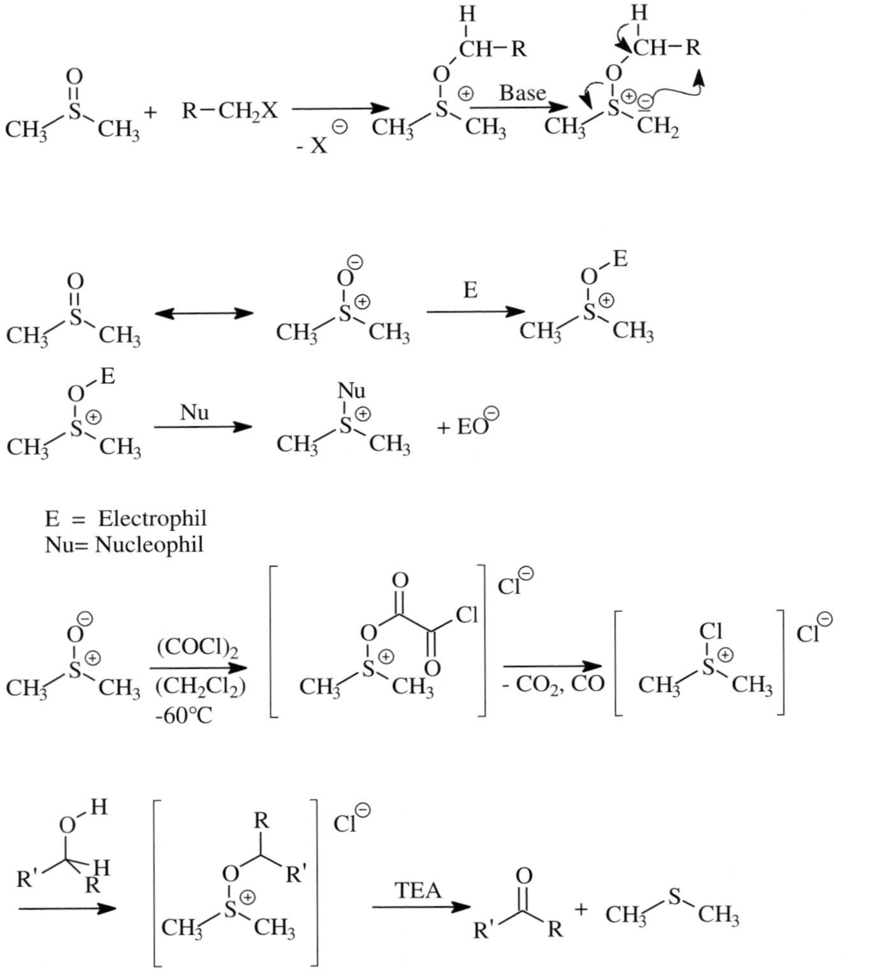

Fig. 5.3. General mechanism of a DMSO-mediated oxidation and the Swern oxidation, the most common type of DMSO-mediated oxidation

The conversion of polysaccharides dissolved in DMSO with carboxylic acid anhydrides using a catalyst is one of the easiest methods for esterification at the laboratory scale. Thus, hydrophobically modified polysaccharides can be achieved reacting starch with propionic anhydride in DMSO, catalysed by DMAP and NaHCO$_3$ [143]. The homogeneous succinoylation of pullulan in DMSO with succinic anhydride in the presence of DMAP as catalyst is another nice example for this approach. Succinoylated pullulan can be synthesised with DS with values up to 1 within 24 h at 40 °C. The dependence of the DS on the ratio succinic anhydride/pullulan is shown in Fig. 5.4. NMR analysis indicated that the carboxylic group is preferably introduced at position 6 [144]. Succinoylation of inulin and dextran can be achieved via a similar procedure [145].

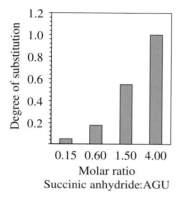

Fig. 5.4. Results (DS determined by titration) for the succinoylation of pullulan with succinic anhydride in DMSO for 24 h at 40 °C (adapted from [144])

For higher aliphatic esters, the use of carboxylic acid halides is necessary. For example, homogeneous esterification of dextran in DMSO with fatty acid halides (C_{10}–C_{14}) for 48 h at 45 °C can be exploited to prepare clearly water-soluble esters with DS values around 0.15 [146].

In addition to the modification of glucanes, DMSO is used as solvent for the homogeneous esterification of the carboxylic acid functions of alginates [5]. The polysaccharide is converted into the acid form, subsequently into the tetrabutylammonium salt by treatment with TBA hydroxide, and finally this salt is converted homogeneously in DMSO with long-chain alkyl bromides (Fig. 5.5).

Modification reactions of glucans, including reagents such as TFAA, oxalyl chloride, TosCl or DCC, should preferably be carried out in formamide, DMF or NMP because conversion in DMSO can be combined with the oxidation at least of the primary OH group to an aldehyde moiety. Side reactions occurring during esterification reactions with DCC in DMSO (e.g. Moffatt oxidation) are discussed in detail in Sect. 5.2.2. Formamide, DMF and NMP can be used as solvent in the same manner as DMSO, e.g. for the acetylation of starch [147]. DMF is used as solvent for the esterification of starch with fatty acids [148]. In addition, synthesis of starch trisuccinate is accomplished in formamide at 70 °C over 48 h using Py as base [149].

A solvent mixture specifically applied for dextrans is NMP/formamide; dextran esters of fatty acids (C_{10}–C_{14}) with DS of 0.005–0.15, soluble in H_2O, can be obtained by conversion with fatty acid halides [146]. More frequently DMF, NMP and DMAc are used in combination with LiCl as solvent.

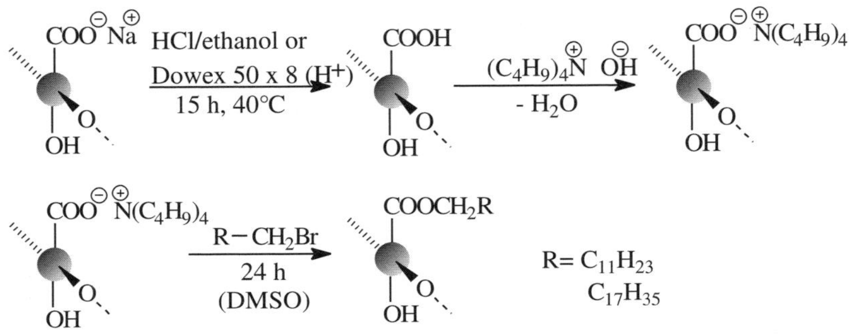

Fig. 5.5. Course of reaction for the esterification of alginate with long-chain alkyl halides (adapted from [5])

Inulin can be dissolved in Py and long-chain fatty acid esters can be prepared homogeneously with the anhydrides, yielding polymers of low DS in the range 0.03–0.06 [107]. For higher functionalisation, the carboxylic acid chloride is used (see Table 4.4).

Alternative single-component solvents used for the esterification of cellulose are organic salt melts, especially N-alkylpyridinium halides. N-ethylpyridinium chloride is extensively studied. The salt melts are often diluted with common organic liquids to give reaction media with appropriate melting points. Among the additives for N-ethylpyridinium chloride (m.p. 118 °C) are DMF, DMSO, sulfolane, Py and NMP, leading to a melting point of 75 °C [150].

Cellulose with DP values up to 6500 can be dissolved in N-ethylpyridinium chloride. The homogeneous acetylation of cellulose in N-ethylpyridinium chloride in the presence of Py is achievable using acetic anhydride, leading to a product with a DS 2.65 within short reaction times of 44 min [123]. Although the preparation of cellulose triacetate, which is completed within 1 h, needs to be carried out at 85 °C, it proceeds without degradation for cellulose with DP values below 1000, i.e. strictly polymeranalogous. Cellulose acetate samples with a defined solubility, e.g. in water, acetone or chloroform, are accessible in one step, in contrast to the heterogeneous conversion (Table 5.5). A correlation between solubility and distribution of substituents has been attempted by means of ^{1}H NMR spectroscopy ([151], see Chap. 8).

Ionic liquids, especially those based on substituted imidazolium ions, are capable of dissolving cellulose over a wide range of DP values (even bacterial cellulose), without covalent interaction (Fig. 5.6, [152]).

Different types of ionic liquids, and the treatment necessary for cellulose dissolution are summarised in Table 5.6. Usually, the polysaccharide dissolves during thermal treatment at 100 °C. The remarkable feature is that acylation of cellulose can be carried out with acetic anhydride in ionic liquids displayed in Fig. 5.6. The reaction succeeds without an additional catalyst. Starting from DS 1.86, the cellulose acetates obtained are acetone soluble [124]. The control of the DS by prolongation of the reaction time is displayed in Table 5.7. When acetyl chlo-

Table 5.5. Preparation of cellulose acetate in N-ethyl-pyridinium chloride (adapted from [132])

| Reaction conditions | | | | | Reaction product | |
| Molar ratio | | | Temp. (°C) | Time (min) | DS | Solubility |
AGU	Py	Acetic anhydride				
1	16.2	5.4	40	60	0.52	H_2O/Py 3/1
1	16.2	5.4	40	295	1.39	CCl_4/methanol 4/1
1	32.5	32.5	50	120	2.25	CCl_4/methanol 4/1
1	32.0	32.0	85	55	2.61	$CHCl_3$
1	32.5	32.5	50	285	2.71	Acetone; $CHCl_3$

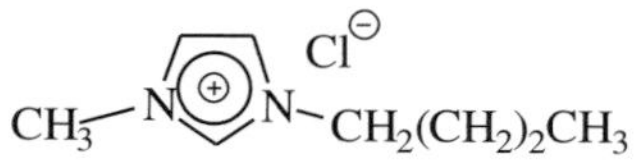

1-N-Butyl-3-methylimid-
azolium chloride, [C$_4$mim]Cl

1-N-Allyl-3-methylimid-
azolium chloride, AMIMCl

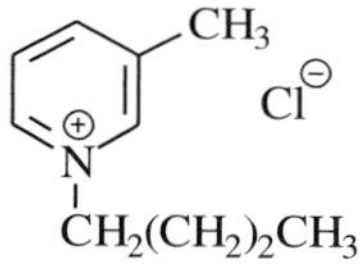

3-Methyl-N-butyl-
pyridinium chloride

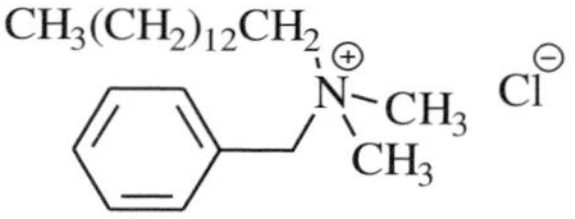

Benzyldimethyl(tetradecyl)-
ammonium chloride

Fig. 5.6. Structures of ionic liquids capable of cellulose dissolution

Table 5.6. Ionic liquids capable of cellulose dissolution (adapted from [152])

Ionic liquid	Method	Solubility (wt%)
[C$_4$mim]Cl	Heat to 100 °C	10
[C$_4$mim]Cl	Heat to 70 °C	3
[C$_4$mim]Cl	Heat to 80 °C, sonication	5
[C$_4$mim]Cl	Microwave treatment	25
[C$_4$mim]Br	Microwave treatment	5–7
[C$_4$mim]SCN	Microwave treatment	5–7
AMIMCl	Heat to 100 °C	5–10

Table 5.7. Acetylation of cellulose in AMIMCl (4%, w/w cellulose, molar ratio AGU:acetic anhydride 1:5, temperature 80 °C, adapted from [124])

Time (h)	DS	Solubility Acetone	Chloroform
0.25	0.94	–	–
1.0	1.61	–	–
3.0	1.86	+	–
8.0	2.49	+	+
23.0	2.74	+	+

ride is added, complete acetylation of cellulose is achieved in 20 min [153]. No other homogeneous acylation experiments are known in this type of solvent; the method may lead to a widely applicable acylation procedure for polysaccharides, if regeneration of the solvent becomes possible.

NMMO, the commercially applied cellulose solvent for spinning (Lyocell® fibres), is usable as medium for the homogeneous acetylation of cellulose with rather low DS values [125]. NMMO monohydrate (about 13% water) dissolves cellulose at ≈ 100 °C. Esterification of dissolved polymer is accomplished in this solvent with vinyl acetate, to give a product with DS 0.3. The application of an enzyme (e.g. Proteinase N of *Bacillus subtilis*) as acetylation catalyst seems to be necessary.

5.1.3 Multicomponent Solvents

The most versatile multicomponent solvent is a mixture of a polar aprotic solvent and a salt. The broadest application was found for the combination substituted amide/LiCl. Most of the glucans discussed above dissolve easily in the mixture DMF/LiCl upon heating to 90–100 °C. Especially in the case of dextran and xylan, this solvent can be exploited for a broad variety of modifications, as displayed in Fig. 5.7 for dextran.

Hydrophobic xylans are accessible homogeneously in DMF/LiCl by conversion under mild reaction conditions with fatty acid chlorides, using TEA/DMAP as base and catalyst (Table 5.8 [157]).

DMAc/LiCl, widely used in peptide and polyamide chemistry, is among the best studied solvents because it dissolves a wide variety of polysaccharides including cellulose, chitin, chitosan, amylose and amylopectin [158]. DMAc/LiCl does not cause degradation, even in the case of high-molecular mass polysaccharides, e.g. potato starch, dextran from *Leuconostoc mesenteroides* or bacterial cellulose. It shows almost no interaction with acylating reagents, and can even act as acylation catalyst.

It is not known how DMAc/LiCl dissolves polysaccharides. A number of solvent-polymer structures for the interaction between cellulose and DMAc/LiCl have been

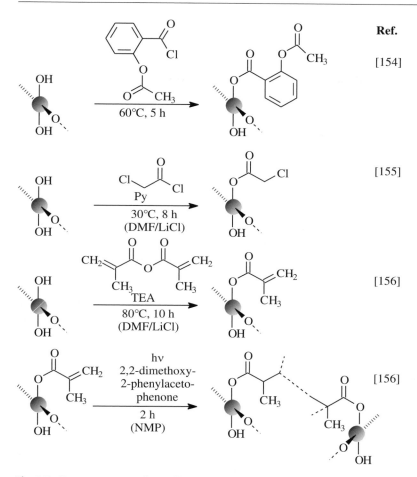

Fig. 5.7. Dextran esters synthesised homogeneously in DMF/LiCl

proposed (Fig. 5.8, [159]). According to [160], the most reasonable structure is the one proposed by McCormick. In addition, the structures of El-Kafrawy and Turbak agree with studies applying solvatochromic polarity parameters, while the structure proposed by Vincendon does not fit in actual results, because Li$^+$ and Cl$^-$ are in contact. The most probable interaction of chitin with the solvent DMAc/LiCl, studied by means of ^{1}H NMR spectroscopy with N-acetyl-D-glucosamine and methyl-D-chitobioside as model compounds, involves a "sandwich-like" structure (Fig. 5.9, [161]).

The dissolution process is rather simple. It can be achieved by solvent exchange, meaning the polysaccharide is initially suspended in water and the polymer is subsequently transferred into methanol and DMAc, i.e. in organic liquids with decreasing polarity, and finally DMAc/LiCl [162]. Dissolution occurs by heating

Table 5.8. Esterification of xylan with acid chlorides

Carboxylic acid chloride	Conditions Molar ratio			Time (min)	Temp. (°C)	Product DS
	AXU	Carboxylic acid chloride	TEA			
Acetyl	1	3	3.2	30	45	0.63
Butyryl	1	3	3.7	35	75	1.15
Octanoyl	1	3	3.7	40	75	1.17
Decanoyl	1	3	2.9	40	75	1.21
Stearoyl	1	2	1.4	30	65	0.40
Stearoyl	1	3	3.7	45	75	1.51
Oleoyl	1	3	2.4	40	75	1.17

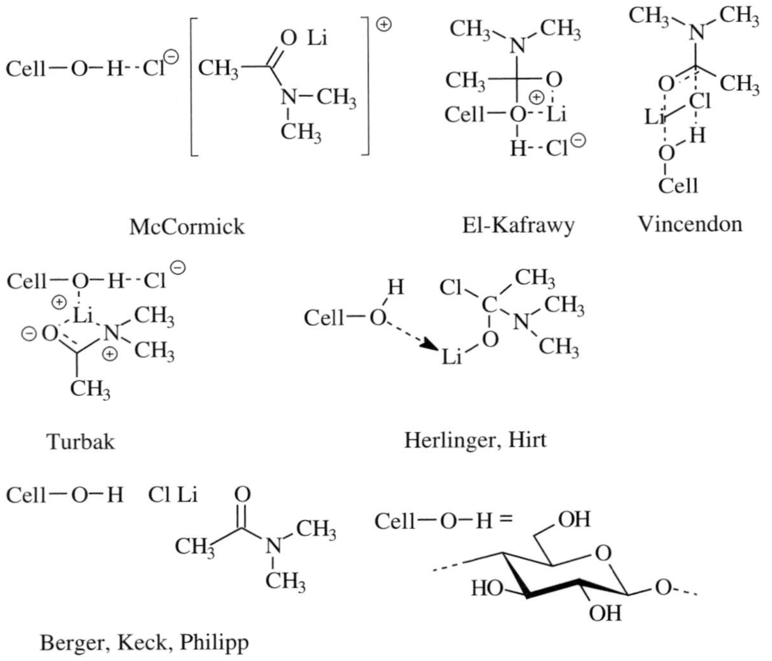

McCormick   El-Kafrawy   Vincendon

Turbak   Herlinger, Hirt

Berger, Keck, Philipp

Fig. 5.8. Proposed solvent structures of cellulose in DMAc/LiCl (adapted from [26] and [160])

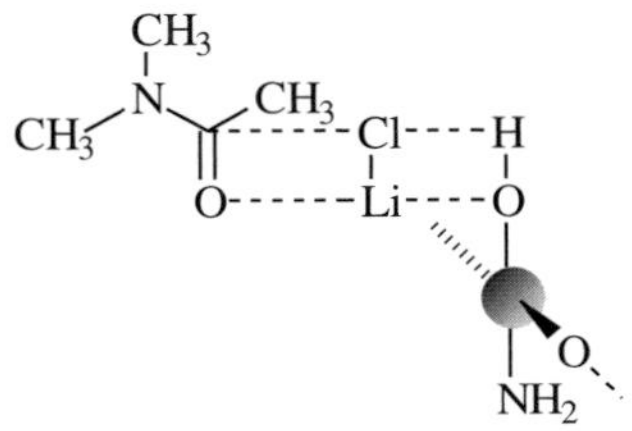

Fig. 5.9. Structure proposed for the interaction of GlcNAc with DMAc/ LiCl as model for the dissolution of chitin

to 80 °C. More commonly used is dissolution after heating a suspension of the polysaccharide in DMAc to 130 °C, evaporating about 1/5 of the liquid (containing most of the water from the polysaccharide) under vacuum, and addition of LiCl at 100 °C. During cooling to room temperature, a clear solution is obtained. The amount of polysaccharide soluble in the mixture varies from 2 to 12% (w/w), depending on the DP of the polysaccharide. The amount of LiCl is in the range 5–15% (w/w). For a standard solution used for chemical modification (see experimental section of this book), 2.5% (w/w) polysaccharide and 7.5% (w/w) LiCl are used.

These solutions are among the most useful tools for the homogeneous synthesis of complex and tailored polysaccharide esters, as described below for the reaction after in situ activation of the carboxylic acid or transesterification reactions. However, conversion of polysaccharides, especially cellulose, in DMAc/LiCl may lead to direct access to cellulose esters that can be processed further (solvent-soluble or melt-flowable). This is due to the high efficiency of the homogeneous reaction conditions and also because acylation without an additional catalyst is possible in this medium, and the solvent system can be recovered almost completely.

In recent years, the cellulose/DMAc/LiCl system has been studied intensively to develop efficient methods appropriate even for industrial application [163, 164]. The dissolution procedure and acetylation conditions in DMAc/LiCl allow excellent control of the DS in the range from 1 to 3. Thermal cellulose activation under reduced pressure is far superior to the costly and time-consuming activation by solvent exchange. Reaction at 110 °C for 4 h without additional base or catalyst gives products with almost no degradation of the starting polymer. A distribution of substituents in the order C-6 > C-2 > C-3 has been determined by means of ^{13}C NMR spectroscopy. In addition to microcrystalline cellulose, cotton, sisal and bagasse-based cellulose may serve as starting material (Table 5.9). The crystallinity of the starting polymer has little effect on the homogeneous acetylation.

Table 5.9. Acetylation of different cellulose types in DMAc/LiCl with acetic anhydride (18 h at 60 °C, adapted from [163] and [164])

Starting materials				Molar ratio		DS
Cellulose from	M_w (g/mol)	α-Cellulose content (%)	I_c (%)	AGU	Acetic anhydride	
Bagasse	116 000	89	67	1	1.5	1.0
Bagasse	116 000	89	67	1	3.0	2.1
Bagasse	116 000	89	67	1	4.5	2.9
Cotton	66 000	92	75	1	1.5	0.9
Sisal	105 000	86	77	1	1.5	1.0

A comparable efficiency is observed for the conversion of cellulose with carboxylic acid chlorides. In the pioneering work of McCormick and Callais [162], acetyl chloride was applied in combination with Py to prepare a cellulose acetate with DS of 2.4, which is soluble in acetone. Detailed information on the DS values attainable, concerning solubility of the acetates and distribution of substituents, are given in Table 5.10.

Table 5.10. Acetylation of cellulose with acetyl chloride in the presence of Py in DMAc/LiCl (adapted from [127])

Reaction conditions			Reaction product			Solubility[a]		
Molar ratio			Partial DS in position					
AGU	Acetyl-chloride	Py	6	2,3	Σ	DMSO	Acetone	CHCl$_3$
1	1.0	1.2	0.63	0.37	1.00	+	−	−
1	3.0	3.6	0.94	1.62	2.56	+	−	+
1	5.0	6.0	0.71	2.0	2.71	+	+	+
1	5.0	10.0	0.46	2.0	2.46	+	+	+
1	4.5	−	1.00	1.94	2.94	+	−	−

[a] + Soluble, − insoluble

^{1}H NMR spectroscopy revealed a comparably high amount of functionalisation at the secondary OH groups [127]. This effect is even more pronounced by an increased concentration of the base. For a sample with an overall DS of 2.46, a partial DS at position 6 of 0.46 is achieved, i.e. all the secondary OH groups are acetylated. This is a first hint for a preferred deacetylation at the position 6 during the reaction. The method yields samples completely soluble in acetone. A rather dramatic depolymerisation of about 60% during the acetylation is concluded from GPC investigations. One possible explanation for the degradation and the pattern of functionalisation might be the formation of the acidic pyridinium hydrochloride in the case of the base-catalysed reaction, causing hydrolysis.

Amazingly, acetylation of cellulose dissolved in DMAc/LiCl with acetyl chloride without an additional base (see Table 5.10) succeeds with almost complete conversion, and can be controlled by stoichiometry. In contrast to the application of Py, higher DS values and a preferred functionalisation of the primary hydroxyl groups are found. Cellulose acetates soluble in acetone are not accessible. Thus, different solubility is due to the different distribution of substituents on the level of the AGU. GPC investigations indicate less pronounced chain degradation during the reaction without a base. In the case of Avicel as starting polymer, depolymerisation is less than 2%. Permethylation, degradation and HPLC do not suggest a non-statistic distribution of the substituents along the polymer chain (see Sect. 8.4.2).

Conversion of glucans with acid chlorides in DMAc/LiCl is most suitable for the homogeneous synthesis of freely soluble, partially functionalised long-chain aliphatic esters and substituted acetic acid esters (Table 5.11). In contrast to the

Table 5.11. Preparation of aliphatic esters of cellulose in DMAc/LiCl

Reaction conditions						Reaction product		
	Molar ratio		Base	Time (h)	Temp. (°C)	DS	Solubility	Ref.
Acid chloride	AGU	Agent						
Hexanoyl	1	1.0	Py	0.5	60	0.89	DMSO, NMP, Py	[99]
Hexanoyl	1	2.0	Py	0.5	60	1.70	Acetone, MEK, CHCl$_3$, AcOH, THF, DMSO, NMP, Py	
Lauroyl	1	2.0	Py	0.5	60	1.83	Py	
Stearoyl	1	1.0	Py	1	105	0.79	Acetone, MEK, CHCl$_3$, AcOH, THF, DMSO, NMP, Py	
Hexanoyl	1	8.0	TEA	8	25	2.8	DMF	[162]
Heptanoyl	1	8.0	TEA	8	25	2.4	Toluene	
Octanoyl	1	8.0	TEA	8	25	2.2	Toluene	
Phenylacetyl	1	15.0	Py	3/1.5	80/120	1.90	CH$_2$Cl$_2$	[165]
4-Methoxy-phenylacetyl	1	15.0	Py	3/1.5	80/120	1.8	CH$_2$Cl$_2$	
4-Tolyl-acetyl	1	15.0	Py	3/1.5	80/120	1.8	CH$_2$Cl$_2$	

anhydrides, the fatty acid chlorides are soluble in the reaction mixture, and very soluble polysaccharide esters may be formed with a very high efficiency of the reaction. Even in the case of stearoyl chloride, 79% of the reagent is consumed for the esterification of cellulose.

Starch and chitin esters can be synthesised in a similar way by homogeneous esterification of the polysaccharides in DMAc/LiCl (Table 5.12). Starch succinates and starch fatty acid esters in almost quantitative yields may be prepared [139, 166].

In addition to the aliphatic esters, a variety of alicyclic, aromatic and heterocyclic esters are accessible, as shown in Fig. 5.10. In addition to DMAc/LiCl, a number of modified compositions of the solvent mixture are known. DMAc can be substituted with NMP, DMF, DMSO, N-methylpyridine or HMPA but only NMP, the cyclic analogue of DMAc, dissolves polysaccharides without major degradation. Furthermore, the mixture of DMI and LiCl is a suitable solvent for cellulose [128]. The advantages of commercially available DMI are its thermal stability and low toxicity. DMI/LiCl is able to dissolve cellulose with DP values as high as 1200 at concentrations of 2–10% (w/w), applying an activation of the polymer by a heat treatment or a stepwise solvent exchange. ^{13}C NMR spectra of cellulose acquired both in DMI and DMAc in combination with LiCl exhibit the same chemical shifts, i.e. comparable cellulose solvent interactions may be assumed. Soluble, partially

Table 5.12. Esterification of chitin and starch homogeneously in DMAc/LiCl using Py as base

Polysaccharide	Carboxylic acid chloride	Conditions				Product	Ref.
		Molar ratio[a]		Temp. (°C)	Time (h)	DS	
		RU	Acid chloride				
Chitin	Acetyl	1	54	110	2	2.0	[167]
	4-Chloro butyroyl	1	50	80	3	2.0	[167]
	Benzoyl	1	48	80	3	1.0	[167]
Starch	Octanoyl	1	1.5	80	0.5	1.5	[166]
	Myristoyl	1	4.5	6	100	2.69	[168]
	Stearoyl	1	4.5	6	100	2.17	[168]
	Palmitoyl	1	1.5	80	0.5	1.5	[166]

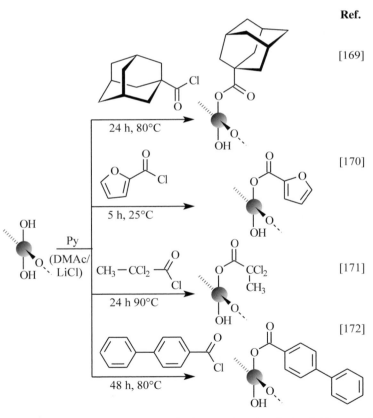

Ref.

[169]

24 h, 80°C

[170]

5 h, 25°C

Py (DMAc/ LiCl)

[171]

24 h 90°C

[172]

48 h, 80°C

Fig. 5.10. Homogeneous synthesis of adamantoyl-, 2-furoyl-, 2,2-dichloropropyl- and 4-phenyl-benzoyl cellulose in DMAc/LiCl

functionalised cellulose acetate (DS 1.4) is obtained by conversion of the polymer with acetic anhydride/Py in DMI/LiCl. β-Ketoesters with DS up to 2.1 were introduced by reaction of cellulose dissolved in DMI/LiCl with cis-9-octadecenyl ketene dimer [173].

DMSO/TBAF is a very useful system, as even cellulose with a DP as high as 650 dissolves without any pre-treatment within 15 min [27]. Highly resolved ^{13}C NMR spectra of cellulose can be obtained showing all the ring carbons of the AGU at 102.7 (C-1), 78.4 (C-4), 75.6 (C-5), 75 (C-3), 73.5 (C-2) and 59.9 ppm (C-6), giving no hints for the formation of covalent bonds during the dissolution process (Fig. 2.1). The solvent is highly efficient as reaction medium for the homogeneous esterification of polysaccharides by transesterification and after in situ activation of complex carboxylic acids (Sect. 5.2).

The acylation using acid chlorides and anhydrides is limited because the solution contains a certain amount of water caused by the use of commercially available TBAF trihydrate and residues of the air-dried polysaccharides. Nevertheless, it has shown a remarkable capacity for the esterification of lignocellulosic materials, e.g. sisal cellulose, which contains about 14% hemicellulose [129]. The DS values of cellulose acetate prepared from sisal with acetic anhydride in mixtures of DMSO/TBAF decrease with increasing TBAF concentration from 6 to 11% (Table 5.13), due to the increased rate of hydrolysis both of the anhydride and also of the ester moieties.

Table 5.13. Influence of the amount of TBAF trihydrate on the efficiency of the acetylation of sisal cellulose with acetic anhydride in DMSO/TBAF (adapted from [129])

%TBAF in DMSO	Cellulose acetate	
	DS	Solubility
11	0.30	Insoluble
8	0.96	DMSO, Py
7	1.07	DMSO, Py
6	1.29	DMSO, DMF, Py

Dehydration of DMSO/TBAF is possible by vacuum distillation; reactions in the solvent of reduced water content lead to products comparable to the reaction of cellulose dissolved in anhydrous DMAc/LiCl. In addition to these basic studies, the conversion of cellulose in DMSO/TBAF with more complex carboxylic acids (e.g. furoyl carboxylic acid) via in situ activation with CDI and the reaction with cyclic compounds such as lactones and N-carboxy-α-amino acid anhydrides can be carried out (Sect. 5.2.3). Although other polysaccharides have not been derivatised in this solvent yet, the mixture should be considered for polysaccharide modification because it is an easily usable tool for laboratory-scale esterification towards pure and highly soluble products.

It is well known that TBAF $\times 3$ H$_2$O is degraded by removing the water yielding [HF$_2$]$^-$ ions [174], which do not dissolve cellulose in combination with DMSO. Quite recently, it was shown that anhydrous TBAF can be obtained by nucleophilic substitution of hexafluorobenzene with cyanide (Fig. 5.11, [175]). The mixture dissolves cellulose, and opens up new horizons for the homogeneous functionalisation of cellulose in DMSO/TBAF [176].

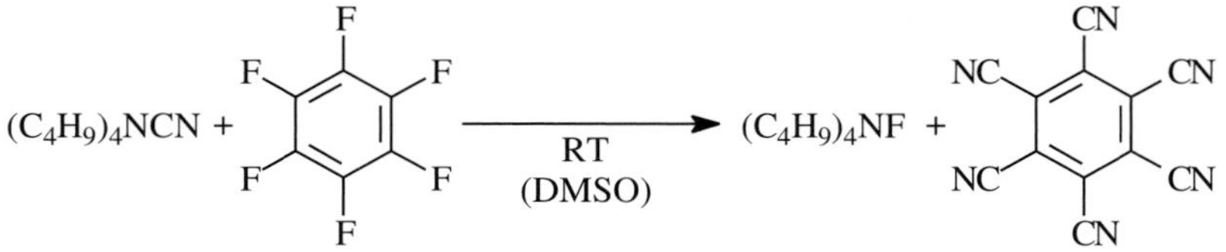

Fig. 5.11. Preparation of anhydrous solvent for cellulose based on DMSO/TBAF

In addition to DMSO/TBAF, mixtures of DMSO with tetraethylammonium chloride can be exploited for the functionalisation of cellulose [176]. For complete dissolution, 25% (w/w) of the salt needs to be added. The cellulose dissolved in this medium is less reactive, compared to DMSO/TBAF system. In addition, mixtures of DMSO with LiCl are utilised for the sulphation of curdlan [177].

5.1.4 Soluble Polysaccharide Intermediates

Formic acid and trifluoroacetic acid are known to dissolve starch [178], guar gum [179], chitin and cellulose [64, 180] at room temperature. Dissolution can be achieved without a co-solvent or a catalyst, depending on the supramolecular structure, the pre-treatment, and the DP. During the dissolution, partial esterification of the polysaccharide occurs and the intermediately formed ester is dissolved. Hence, these solvents are referred to as derivatising solvents. ^{13}C NMR spectroscopy shows that the esterification proceeds preferentially at the primary OH groups. Consequently, esterification to the goal structure is more pronounced at the secondary hydroxyl functions.

Solutions of cellulose (regenerated cellulose, rayon, cellophane) in a surplus of formic acid are obtained without catalyst over periods of 4–15 days [180]. The dissolution is much faster in the presence of sulphuric acid as catalyst. The treatment yields fairly degraded polymers. In contrast, an even faster dissolution of amylose and purified guar gum (rather completely, within 24 h) in formic acid (90% w/w) is observed [181]. Solutions of starch in formic acid can be used directly for synthesis of long-chain starch esters (C$_8$–C$_{18}$), applying fatty acid chlorides in the presence of Py (Table 5.14, [182]).

Cellulose dissolved in TFA can also be used for the acylation of the polysaccharide with carboxylic acid anhydrides or chlorides [183]. An interesting approach for this homogeneous acylation of cellulose in TFA is the treatment of the polymer with carboxylic acids (C$_2$ to C$_9$) in the presence of acetic anhydride [184].

Table 5.14. Fatty acid esters of starch obtained by conversion in formic acid applying molar ratios of 1:6:4.3 (mol AGU/mol fatty acid chloride/mol Py) at 105 °C for 40 min (adapted from [182])

Fatty acid moiety	DS
Octanoate	1.7
Caprinate	1.6
Laurate	1.7
Myristate	1.3
Palmitate	1.1
Stereate	0.8

The reagents are the mixed anhydrides formed intermediately. Highly soluble, completely functionalised mixed esters can be synthesised (Table 5.15).

Table 5.15. Preparation of mixed cellulose esters in TFA, using mixtures of acetic anhydride/carboxylic acid (adapted from [184])

Reaction conditions				Reaction product			
Carboxylic acid	Molar ratio			$DS_{Acetate}$	DS_{Acyl}	Solubility	
	Acetic anhydride	Carboxylic acid	Time (min)			Acetone	Benzene
Propionic	4.5	4.5	11	1.39	1.58	+	+
Propionic	4.5	9.0	20	0.99	2.00	+	+
Propionic	4.5	13.5	35	0.80	2.20	+	+
Butyric	4.5	4.5	11	1.37	1.60	+	+
Butyric	4.5	9.0	22	0.96	2.04	+	+
Butyric	4.5	13.5	35	0.75	2.25	+	+
Caproic	4.5	4.5	12	1.31	1.64	+	+
Caproic	4.5	9.0	30	1.03	1.97	+	−
Caproic	4.5	13.5	52	0.88	2.12	+	−

In the case of cellulose and starch, the isolation of the intermediately formed esters (or transient derivatives) and their subsequent esterification in an inert organic solvent can be carried out. DMSO-, DMF- and Py-soluble cellulose formates with DS values up to 2.5 are attainable in formic acid with sulphuric acid as catalyst (see Chap. 12) or if partially hydrolysed $POCl_3$ is applied as swelling and dehydrating agent [185–187].

Pure cellulose trifluoroacetates (DS 1.5), soluble in DMSO, Py and DMF, can be easily prepared by treating cellulose with mixtures of TFA and TFAA [188]. Formates of starch and amylose are formed simply by dissolving corn starch or amylose in 90% formic acid (1 g in 10 ml) for 2–5 h and precipitation in methanol [178]. The polysaccharide intermediates show preferred functionalisation of the primary OH moiety, as revealed by ^{13}C NMR spectroscopy (shown for cellulose trifluoroacetate in Fig. 5.12).

The subsequent functionalisation of the polysaccharide intermediates with carboxylic acid chlorides homogeneously using DMF as solvent yields products with an interesting pattern of substitution. The conversion of CTFA (DS 1.50) with carboxylic acid chlorides for 4 h at 40 °C and precipitation in water gives soluble and pure (no trifluoroacetyl groups) cellulose esters (Table 5.16).

Table 5.16. Esterification of CTFA in DMF with acid chlorides and [a]via in situ activation with TosCl (see Sect. 5.2)

Carboxylic acid chloride	DS	Solubility
4-Nitro-cinnamic	0.23	DMSO
4-Nitro-benzoic	0.14	DMSO
4-Nitro-benzoic[a]	0.76	DMSO
Palmitic	0.51	DMF

Even the preparation of unsaturated esters, e.g. cinnamates or acrylates with DS values as high as 2.0, is possible [183]. The free acids in combination with TFAA or the acid chlorides can be utilised. Moreover, the homogeneous acetylation of starch in formic acid is an interesting approach towards starch acetates with DS values up to 2.2. Products with an uncommon distribution of substituents are formed (70% of the primary OH groups are not acetylated [178]).

By application of modern organic reagents, e.g. CDI under aprotic conditions, subsequent functionalisation of intermediates gives final products with inverse patterns of functionalisation to that of the starting intermediate, with negligible side reactions, i.e. the primary substituent acts as a protective group and is usually simply cleaved off during the workup procedure under protic conditions, e.g. precipitation in water. This is shown in Fig. 5.12 for the nitrobenzoylation of cellulose, starting from CTFA. The ^{13}C NMR spectrum of the nitrobenzoate (lower part of the figure) shows the preferred substitution of the secondary OH groups. CTFA is the most promising intermediate because of its simple preparation, combined with the highest DP attainable (CTFA with DP values of up to 820 are obtained), its solubility in a wide variety of common organic solvents, its stability under aprotic conditions, and its fast cleavage under aqueous conditions.

Derivatising solvents are summarised, including the intermediates formed by interaction with the polysaccharides, mainly cellulose, in Table 5.17. The major disadvantage of the derivatising solvents is the occurrence of side reactions during dissolution, and the formation of undefined structures leading to products that are hardly reproducible. Accordingly, the intermediate introduction of a primary substituent may lead to a new pattern of substitution, as discussed for the formates and trifluoroacetates of cellulose, starch and guar gum.

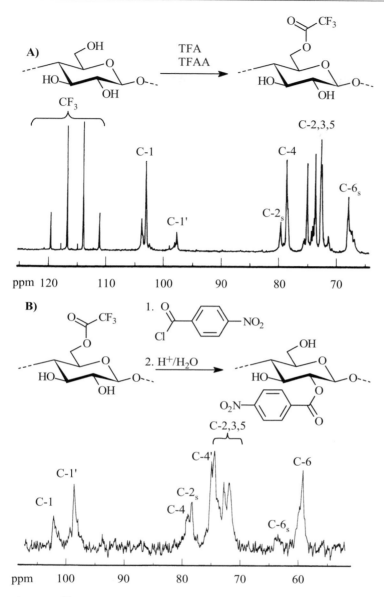

Fig. 5.12. ^{13}C NMR spectrum of **A** cellulose trifluoroacetate (DS 1.50, reprinted from Cellulose 1, Readily hydrolyzable cellulose esters as intermediates for the regioselective derivatization of cellulose; 2, Soluble, highly substituted cellulose trifluoroacetates, pp 249–258, copyright (1994) with permission from Springer) and **B** cellulose nitrobenzoate (DS 0.76) obtained by subsequent esterification, showing inverse patterns of functionalisation

Despite toxicity, DMF/N_2O_4 has found considerable interest in the synthesis of inorganic cellulose esters, e.g. cellulose sulphuric acid half esters and cellulose acetates [192]. Dissolution occurs by yielding cellulose nitrite as intermediate. Instead of DMF, DMSO can be used with N_2O_4, nitrosyl chloride, nitrosyl sulphuric acid, nitrosyl hexachloroantimonate or nitrosyl tetrafluoroborate, forming solutions of polysaccharide nitrite.

Table 5.17. Derivatising solvents applied for cellulose acetylation

Solvent	Intermediate formed	Acetylating reagent	DS_{max}	Ref.
N_2O_4/DMF	Cellulose nitrite	Acetic anhydride	2.0	[189]
Paraformaldehyde/DMSO	Methylol cellulose	Acetic anhydride		[190]
		Acetyl chloride		
		Ethylene diacetate	2.0	
Chloral/DMF/Py	Cellulose trichloroacetal	Acetic anhydride	2.5	[191]

A rather interesting derivatising solvent utilised for esterification is the mixture DMSO/paraformaldehyde, which dissolves cellulose rapidly and almost without degradation, even in the case of a high molecular mass. The polysaccharides are dissolved by formation of the hemiacetal, i.e. the so-called methylol polysaccharide is obtained (Fig. 5.13, [193, 194]). In addition, during the dissolution oligooxy methylene chain formation may occur.

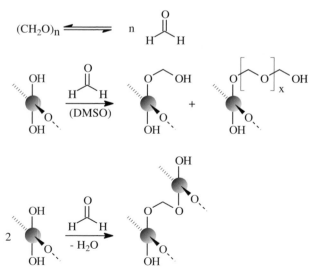

Fig. 5.13. Structure of methylol derivatives formed by dissolution of polysaccharides in DMSO/paraformaldehyde (adapted from [193])

^{13}C NMR spectroscopy shows that the acetalisation occurs preferentially at the position 6 of the AGU of cellulose. This methylol structure remains intact during subsequent functionalisation in non-aqueous media, resulting in derivatives with a pronounced substitution of the secondary OH groups, as can be determined by means of GLC after complete hydrolysis of the subsequently etherified cellulose. The methylol functions can be easily removed by a treatment with water. In addition to the methylol functions, the free terminal hydroxyl groups of the oligooxy methylene chains may also be derivatised in a subsequent step. Nevertheless, DMSO/paraformaldehyde is exploited for the synthesis of esters via homogeneous conversion with a number of carboxylic acid anhydrides including trimellitic anhydride, trimethyl acetic anhydride and phthalic anhydride in the presence of Py [195]. DS values are usually in the range 0.2–2.0, except in the case of acetylation where DS values of up to 2.5 are attainable. Besides DMSO/paraformaldehyde, DMF and DMAc can be used as solvent in combination with paraformaldehyde.

Cellulose dissolves in the mixture chloral/DMF/Py by substitution of the hydroxyl groups to the corresponding hemiacetal groups, which can be acetylated to give products with DS of 2.5 by acetic anhydride or acetyl chloride [191].

5.2 In Situ Activation of Carboxylic Acids

The synthetic approach of in situ activation of carboxylic acids is characterised by reacting the carboxylic acid with a reagent leading to an intermediately formed, highly reactive carboxylic acid derivative. The carboxylic acid derivative may be formed prior to the reaction with the polysaccharide or converted directly in a one-pot reaction. Usually, these reactions are carried out under completely homogeneous conditions. Therefore, the application of an "impeller", which is basically one of the oldest attempts in this regard (reaction via mixed anhydrides, see Chap. 4), is not discussed in this context.

The modification of polysaccharides with carboxylic acids after in situ activation has made a broad variety of new esters accessible because, for numerous acids, e.g. unsaturated or hydrolytically instable ones, reactive derivatives such as anhydrides or chlorides can not be simply synthesised. The mild reaction conditions applied for the in situ activation guard against common side reactions such as pericyclic reactions, hydrolysis, and oxidation. Moreover, due to their hydrophobic character, numerous anhydrides are not soluble in organic media used for polysaccharide modification, resulting in unsatisfactory yields and insoluble products. In addition, the conversion of an anhydride is combined with the loss of half of the acid during the reaction. Consequently, in situ activation is much more efficient. In this chapter, general procedures, model reactions elucidating the reaction mechanisms, and selected examples illustrating the potential of these methods are described.

5.2.1 Sulphonic Acid Chlorides

One of the early attempts for in situ activation is the reaction of carboxylic acids with sulphonic acid chlorides, and the conversion of the acid derivative formed with a polysaccharide. Esterification was accomplished under heterogeneous conditions by conversion of the polymer suspended in Py or DMF with acetic acid, higher aliphatic acids and benzoic acid, using TosCl or MesCl, yielding esters with a wide range of DS values [196, 197]. Similarly, C_{11}–C_{18} acid esters of cellulose can also be obtained [198].

In situ activation using sulphonic acid chlorides has been adopted for the homogeneous modification of polysaccharides, most commonly in DMF/LiCl or DMAc/LiCl. For the majority of the reactions, a base is used although, on the basis of our experiences, this is not necessary [199]. The exclusion of the base simplifies the reaction medium and the isolation procedure. Products synthesised by this route should be purified carefully by precipitation in ethanol or isopropanol, reprecipitation, and Soxhlet extraction.

There is an ongoing discussion about the mechanism that initiates esterification of polysaccharides with carboxylic acids in the presence of TosCl. The mixed anhydride of TosOH and the carboxylic acid is favoured [200, 201]. In contrast, from ^{1}H NMR experiments of acetic acid/TosCl, it can be concluded that a mixture of acetic anhydride (2.21 ppm) and acetyl chloride (2.73 ppm) is responsible for the high reactivity of this system (Figs. 5.14 and 5.15).

By the in situ activation using sulphonic acid chlorides, covalent binding of bioactive molecules onto dextran was achieved by direct esterification of the polymer with α-naphthylacetic acid, naproxen and nicotinic acid homogeneously (DMF/LiCl) using TosCl or MesCl and Py in 22 h at 30–70 °C (Fig. 5.16).

The reaction is influenced by the temperature, Py concentration, and sulphonic acid chloride (Table 5.18, [202]). The esterification is possible even without the base.

^{13}C NMR spectra of partially modified dextran with α-naphthylacetate moieties show that the reactivity of the individual hydroxyl groups decreases in the order C-2 > C-4 > C-3. This distribution is comparable with the one obtained for the acetylation of dextran with acetyl chloride/Py [202]. On the basis of these results, a mechanism for the reaction is suggested, which includes formation of an acylium salt, as observed for the reaction with acid chlorides (Fig. 5.17). These findings support the NMR results discussed above, i.e. the in situ activation with sulphonic acid chloride succeeds mainly via the intermediately formed acyl chlorides of carboxylic acids [202]. The introduction of N-acylamino acid into the dextran backbone can be achieved in the same manner [203].

Cellulose esters, having alkyl substituents in the range from C_{12} (laurylic acid) to C_{20} (eicosanoic acid), can be obtained with almost complete functionalisation of the OH groups within 24 h at 50 °C in the presence of sulphonic acid chlorides, using Py as base (DS values 2.8–2.9, [204]) in DMAc/LiCl. This is also a general method for the in situ activation of waxy carboxylic acids.

The reaction proceeds in 4 h to give partially functionalised fatty acid esters of maximum DS (see entries 1, 7, 9 in Table 5.19). The addition of an extra base

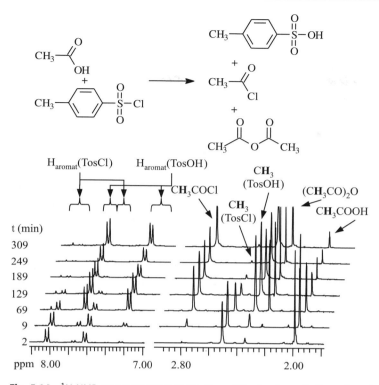

Fig. 5.14. ^{1}H NMR spectroscopic investigation of the in situ activation of acetic acid with TosCl, showing the preferred formation of acetic anhydride and acetyl chloride

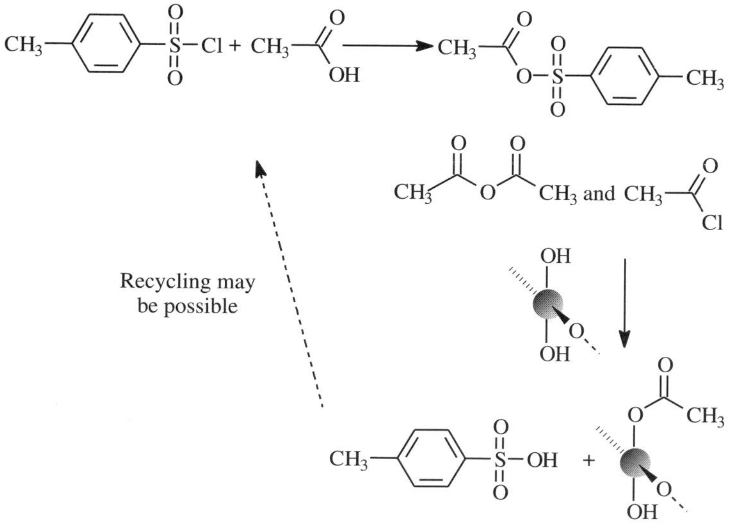

Fig. 5.15. Schematic plot of the reaction involving in situ activation of carboxylic acids with TosCl

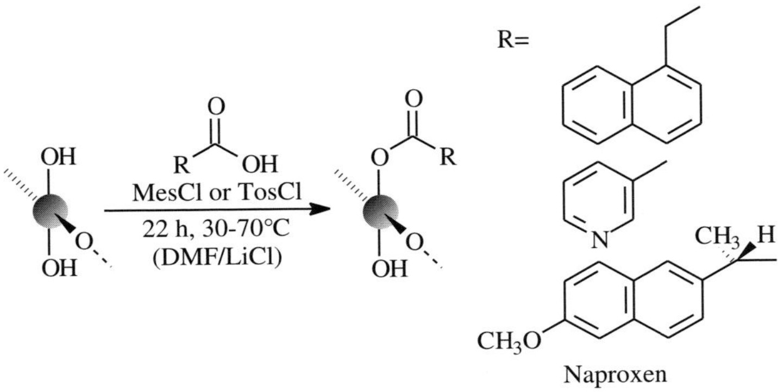

Fig. 5.16. Structures of α-naphthylacetic acid-, nicotinic acid- and naproxen esters of dextran (adapted from [202])

Table 5.18. Esterification of dextran (0.12 mol/l) in DMF/LiCl with α-naphthylacetic acid (0.37 mol/l) in the presence of sulphonic acid chlorides and Py for 22 h (adapted from [202])

Conditions			Product
[Py] (mol/l)	Temp. (°C)	Sulphonic acid chloride	DS
–	50	TosCl	0.13
0.37	50	TosCl	0.15
0.74	50	TosCl	0.19
1.48	50	TosCl	0.23
0.74	30	TosCl	0.13
0.74	60	TosCl	0.22
0.74	70	TosCl	0.23
0.74	50	MesCl	0.18

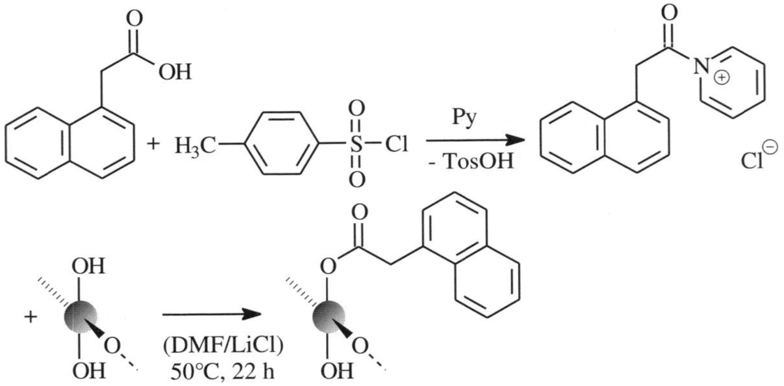

Fig. 5.17. Possible reaction path proposed for the conversion of dextran with α-naphthylacetic acid using TosCl in the presence of Py (adapted from [202])

Table 5.19. Esterification of cellulose dissolved in DMAc/LiCl mediated by TosCl with different carboxylic acids (adapted from [127])

Entry	Reaction conditions						Product	
	Molar ratio						DS	Solubility in CHCl$_3$
	Carboxylic acid	AGU	Acid	TosCl	Py	Time (h)		
1	Lauric	1	2	2	0	24	1.55	+
2	Palmitic	1	2	2	0	24	1.60	+
3	Stearic	1	2	2	0	24	1.76	+
4	Lauric	1	2	2	4	24	1.79	+
5	Palmitic	1	2	2	4	24	1.71	+
6	Stearic	1	2	2	4	24	1.92	+
7	Lauric	1	2	2	0	4	1.55	+
8	Palmitic	1	2	2	0	4	1.50	+
9	Lauric	1	2	2	0	1	1.36	−
10	Palmitic	1	2	2	0	1	1.36	+

increases the DS only in the range of 0.1–0.2 DS units (see entries **1–3** and **4–6** in Table 5.19, [127]). The solubility and the thermal stability of the products are comparable.

Starch can also be esterified with long-chain aliphatic acids activated with TosCl in DMAc/LiCl. The reactions proceed with higher conversions in the presence of CDI, which also decreases the yield of degradation products.

In situ activation with TosCl is also applicable for the introduction of fluorine-containing substituents, e.g. 2,2-difluoroethoxy-, 2,2,2-trifluoroethoxy- and 2,2,3,3,4,4,5,5-octafluoropentoxy functions (synthesised from the fluorinated alcohols with monochloroacetic acid) with DS values mainly in the range 1.0–1.5, leading to a stepwise increase of the hydrophobicity of the products and an increased thermal stability (Fig. 5.18, [201, 205, 206]). Structure analysis is possible by ^{19}F NMR spectroscopy (Fig. 5.19).

Moreover, in situ activation with TosCl enables the synthesis of water-soluble cellulose esters by the derivatisation of cellulose with oxacarboxylic acids in DMAc/LiCl [199]. The conversion of the polysaccharide with 3,6,9-trioxadecanoic acid or 3,6-dioxaheptanoic acid in the presence of TosCl yields non-ionic cellulose esters with DS values in the range 0.4–3.0 (Table 5.20). In this case, the esterification is carried out without an additional base.

A typical ^{13}C NMR spectrum of a cellulose 3,6,9-trioxadecanoic acid ester is shown in Fig. 5.20. No undesired side products are evident. The cellulose derivatives start to dissolve in water at a DS as low as 0.4. They are soluble in common organic solvents such as acetone or ethanol, and thermally stable up to 325 °C.

Cellulose-2,2-difluoro-ethoxyacetate

Cellulose-2,2,3,3,4,4,5,5-octafluoropentoxyacetate

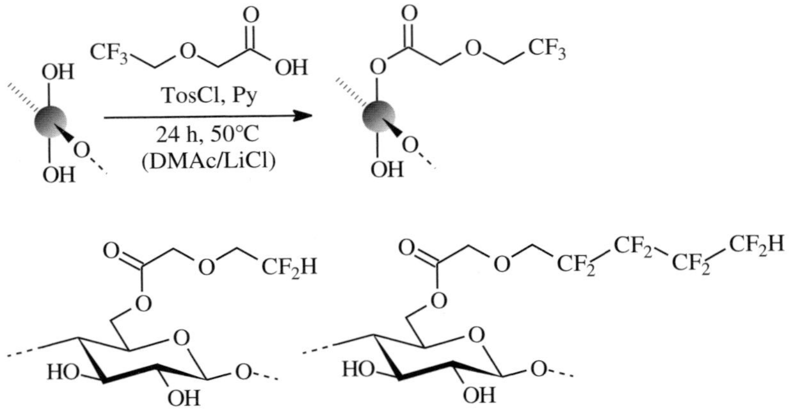

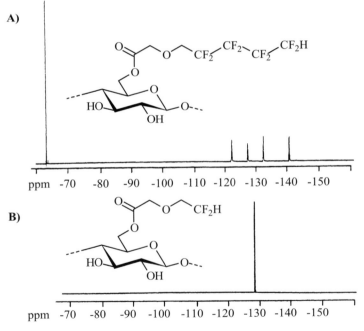

Fig. 5.18. Structures of fluorine-containing cellulose derivatives attainable via in situ activation of the carboxylic acid with TosCl (adapted from [206])

Fig. 5.19. ^{19}F NMR spectra of cellulose 2,2,3,3,4,4,5,5-octafluoropentoxyacetate (**A**, signal at −64.1 ppm is caused by the standard 3-(trifluoro)methyl benzophenone) and cellulose difluoroethoxy acetate (**B**, DS 1.0) (reprinted from Carbohydr Polym 42, Glasser et al., Novel cellulose derivatives. Part VI. Preparation and thermal analysis of two novel cellulose esters with fluorine-containing substituents, pp 393–400, copyright (2000) with permission from Elsevier)

Table 5.20. Esterification of cellulose with 3,6,9-trioxadecanoic acid (TODA) or 3,6-dioxahexanoic acid (DOHA) mediated with TosCl (1 equivalent acid) for 3 h at 65 °C (adapted from [199])

Conditions			Product			
Acid	Molar ratio		DS	Solubility		
	AGU	Acid		Water	Acetone	Ethanol
TODA	1	1.8	0.43	+	–	–
TODA	1	3.0	0.62	+	+	–
TODA	1	9.0	3.00	+	+	+
DOHA	1	3.0	1.66	+	+	–
DOHA	1	6.0	1.87	+	+	–

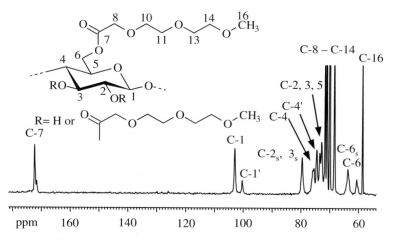

Fig. 5.20. ^{13}C NMR spectrum of a cellulose 3,6,9-trioxadecanoic acid ester (DS 1.18, reproduced with permission from [199], copyright Wiley VCH)

Even cellulose derivatives with bulky substituents and with DS values as high as 1.0 are obtainable via in situ activation of carboxylic acids with TosCl. Fluorescence-active cellulose anthracene-9-carboxylate can be prepared [207].

Homogeneous acylation of pullulan in DMSO with abietic acid/TosCl yields low-substituted pullulan abietates (DS up to 0.4) usable for the surface modification of cellulose (Langmuir Blodgett films) to biomimetic wood composites [208].

In summary, the application of a sulphonic acid chloride, especially TosCl, for the in situ activation of carboxylic acids is an easy procedure, valuable for the preparation of long-chain aliphatic and alicyclic esters of polysaccharides. If DMF or DMAc is used, then an extra base is not required. The reactions are accompanied by pronounced chain degradation. Careful removal of by-products is necessary, preferably by extraction with ethanol, and should be controlled by sulphur analysis.

In addition to the sulphonic acid chlorides, application of TosOH as catalyst for the conversion of curdlan with acetic anhydride in acetic acid has been studied. This is a catalytic process, and not an in situ activation [209].

The reaction with the alkali or alkaline earth salt of acetic acid with TosCl as catalyst has also been reported [197].

5.2.2 Dialkylcarbodiimide

Coupling reagents of the dialkylcarbodiimide type are most frequently utilised for the esterification of polysaccharides with complex carboxylic acids. The most widely used, in particular in peptide and protein chemistry, is DCC (Fig. 5.21, [210]).

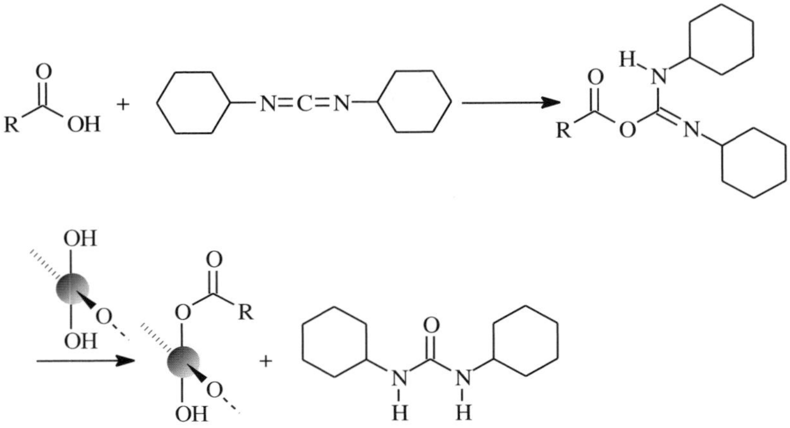

Fig. 5.21. Esterification of a polysaccharide with carboxylic acid in situ activated with DCC (adapted from [210])

These reagents have a number of drawbacks. First of all, they are toxic, particularly upon contact with the skin. The LD_{50} (dermal, rat) of DCC is 71 mg/kg. This should always be considered if the reaction is used for the preparation of materials for biological applications. Moreover, the N,N-dialkylurea formed during the reaction is hard to remove from the polymer, except for preparation in DMF and DMSO where it can be filtered off. In the case of esterification of polysaccharides in DMSO in the presence of these reagents, oxidation of hydroxyl functions may occur due to a Moffatt-type reaction (Fig. 5.22, [211]).

The oxidation products formed can be detected with the aid of 2,4-dinitrophenylhydrazine, e.g. in the case of the conversion of dextran with DCC in DMSO [212]. Moreover, during the reaction, decomposition of DMSO to dimethylsulphide occurs, resulting in a pungent odour. The treatment with DCC may also lead to the formation of isourea ethers, according to the reaction shown in Fig. 5.23.

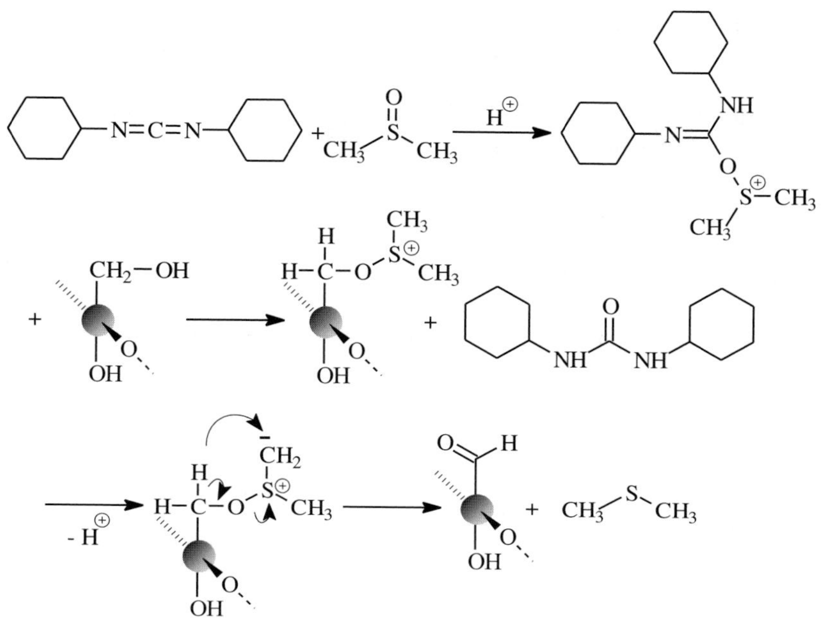

Fig. 5.22. Mechanism for the Moffatt oxidation of the primary OH group of a polysaccharide with DCC (adapted from [211])

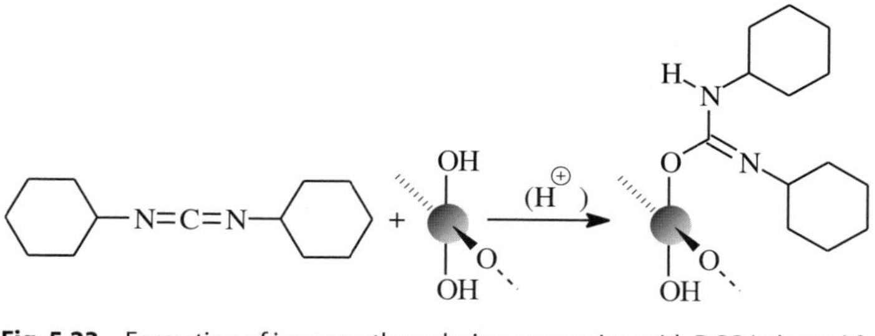

Fig. 5.23. Formation of isourea ethers during conversion with DCC (adapted from [212])

Despite these problems, a number of esterification reactions have been described using DCC. For most of the derivatives discussed, structure analysis is limited to the determination of the percentage of bound acid. Moreover, the type of binding (formation of side chains, MS) is not considered and analysed. In any case, the side reactions mentioned should be taken into account and evaluated. A critical discussion of the use of DCC in comparison with CDI can be found in Sect. 5.2.3. According to our own experiences, CDI should always be considered as more appropriate for the introduction of complex ester moieties.

DCC is most frequently used in combination with DMAP as a catalyst, and a number of sophisticated polysaccharide esters are accessible. Although widely exploited, the efficiency of the reaction is usually rather low, i.e. the acids are converted with 10–30% yield only. Another, more efficient approach is the conversion with the mixed reagent DCC/PP, as discussed for the introduction of long-chain fatty acids [99]. A solvent mixture useful for the acylation of polysaccharides such as dextran and pullulan is formamide/DMF/CH_2Cl_2.

DCC is used for the covalent binding of numerous biomolecules onto polysaccharides. It is applied for the introduction of protected amino acids. The fructan inulin can be modified by reaction with N,N'-bis-benzyloxycarbonyl-L-lysine and N,N-benzyloxycarbonylglycine using DCC/DMAP (Fig. 5.24, [213]). The conversion of inulin dissolved in DMF succeeds at very mild reaction conditions (RT, 6 h). The inulin–lysin has a DS of 0.95 and the inulin–glycin a DS value of 1.01. The resulting polymers can be deprotected by the catalytic transfer hydrogenation method, using 1,4-cyclohexadiene as an effective hydrogen donor.

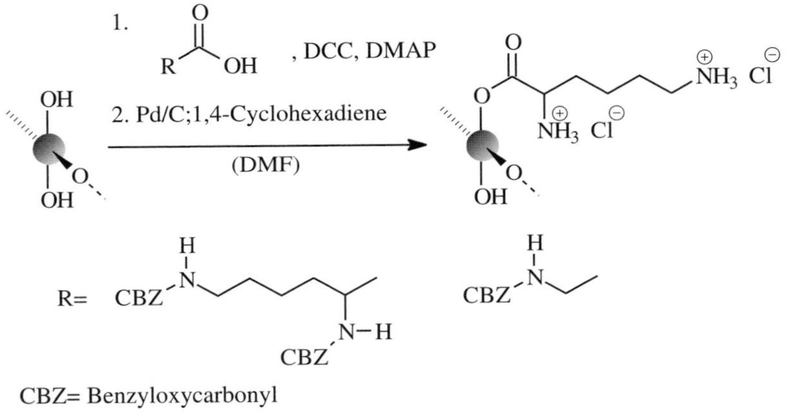

Fig. 5.24. Synthesis path for inulin amino acid esters synthesized with N,N'-bis-benzyloxycarbonyl-L-lysine and N,N-benzyloxycarbonylglycine, using DCC/DMAP (adapted from [213])

The synthesis of dextran amino acid esters has been achieved by conversion of the polysaccharide in DMSO with the N-benzyloxycarbonyl protected acids for 48 h at 20 °C, using DCC and Py. O-(N-Benzyloxycarbonylglycyl)dextran with DS 1.1, O-(N-benzyloxycarbonylaminoenanthyl)dextran with DS 2.2 and O-(N-acetyl-L-histidinyl)dextran with DS 1.1 are obtainable. Deprotection is achieved with oxalic acid and Pd/C [214, 215].

Functionalisation with bulky hydrophobic carboxylic acids/DCC was studied for the synthesis of amphiphilic polymers based on dextran and pullulan. Bile acid is covalently bound to dextran (Fig. 5.25) through an ester linkage in the presence of DCC/DMAP (added in dichloromethane) as the coupling reagent. The process is homogeneous in DMF/formamide. The amount of bound acid (determined by UV/Vis spectroscopy) is in the range of 10.8 to 11.4 mol% [216, 217].

Subsequent esterification of hydrophobic pullulan acetate with carboxymethy-lated poly(ethylene glycol) applying DCC/DMAP in DMSO leads to amphiphilic polymers (Fig. 5.25). The N,N'-dicyclohexylurea formed during the reaction is removed by filtration. Approximately 42% carboxymethylated poly(ethylene gly-col) is bound to the pullulan backbone [218]. Hydrophobically modified pullulan is achieved by direct esterification of pullulan in DMSO with a perfluoroalkyl carboxylic acid (e.g. $C_8F_{17}CH_2CH_2COOH$) in the presence of DCC/DMAP for 24 h at 40 °C. The content of hydrophobic groups is rather low (1.1–4.8%) [219]. Hydrophobisation of pullulan with a low degree of functionalisation is also accom-plished by reaction with adenine-9-butyric acid and thymine-1-yl butyric acid (Table 5.21, [220]).

Table 5.21. Hydrophobisation of pullulan with adenine-9-butyric acid and thymine-1-yl butyric in DMSO at RT for 48 h with DCC/DMAP (adapted from [220])

Conditions			Product
Reagent	Molar ratio		DS
	AGU	Acid	
Adenine-9-butyric acid	1	0.50	0.05
Adenine-9-butyric acid	1	0.75	0.06
Adenine-9-butyric acid	1	1.20	0.07
Thymine-1-yl butyric acid	1	0.50	0.05

More exotic compounds (prodrugs, see Chap. 10) that are accessible via in situ activation of the carboxylic acids with DCC are the ester of carboxylated sulphonylurea and pullulan [221] synthesised in DMSO, which shows an influence on the insulin secretion, and metronidazole monosuccinate esters of dextran and inulin synthesised in DMSO or DMF, using TEA as base (Fig. 5.26, [222]).

Another strategy for in situ activation is the exploitation of DCC in combi-nation with 4-pyrrolidinopyridine usable for the synthesis of aliphatic cellulose esters [223]. Among the advantages of the method are the high reactivity of the intermediately formed mixed anhydride with PP, and a completely homogeneous reaction in DMAc/LiCl up to hexanoic acid. If the reaction is carried out with the anhydrides of the carboxylic acids, then the carboxylic acid liberated is recycled by forming the mixed anhydride with PP, which is applied only in catalytic amounts. The toxic DCC can be recycled from the reaction mixture (Fig. 5.27). In addition, the method is utilised to obtain unsaturated esters (e.g. methacrylic-, cinnamic- and vinyl acetic acid esters) and esters of aromatic carboxylic acids including (p–N,N-dimethylamino)benzoate of cellulose. The amino group-containing ester provides a site for the conversion to a quaternary ammonium derivative, which imparts water solubility [224, 225].

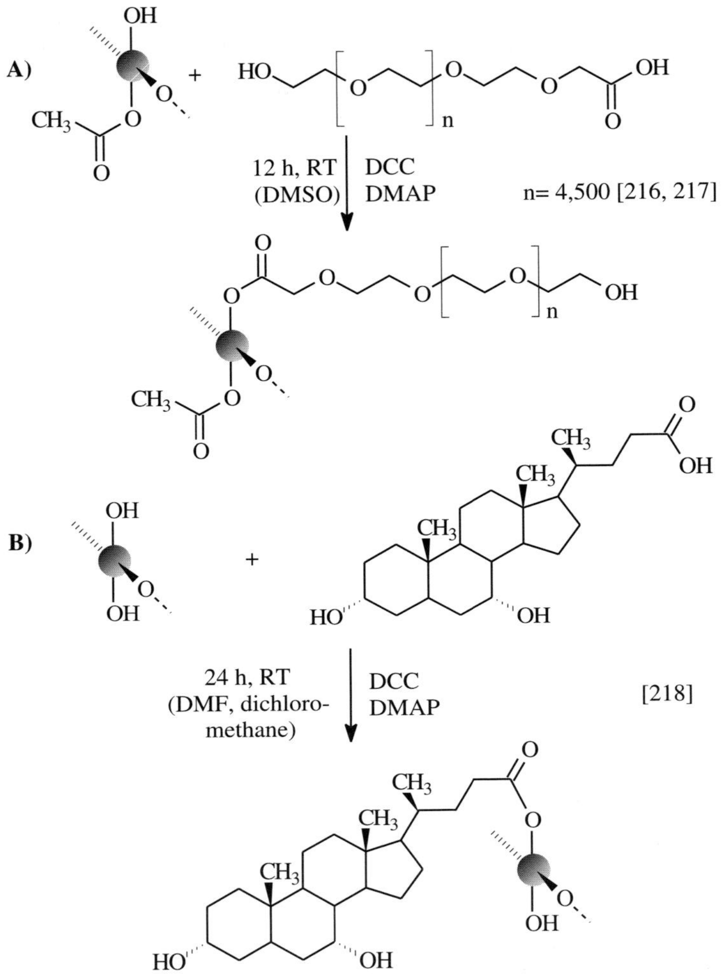

Fig. 5.25. Structures of amphiphilic polymers **A** ester of carboxymethylated poly(ethylene glycol) with pullulan acetate and **B** bile acid ester of dextran prepared by esterification via in situ activation with DCC/DMAP

An alternative to DCC is 1-ethyl-3-(3-dimethylaminopropyl)carbodiimide, which is used for the hydrophobic modification of pullulan [226], and diisopropylcarbodiimide, which can be utilised to introduce betaine moieties into starch [227].

5.2.3 *N,N'*-Carbonyldiimidazole

A method with an enormous potential for polysaccharide modification is the homogeneous one-pot synthesis after in situ activation of the carboxylic acids with CDI, which is rather well known from organic chemistry literature published

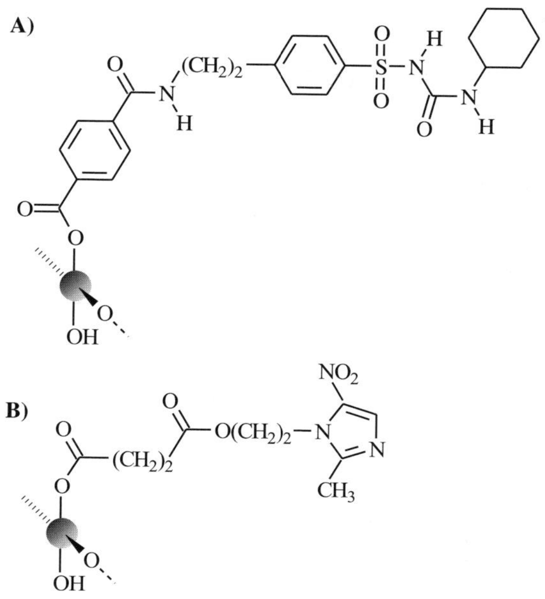

A)

Fig. 5.26. Structure of the esters of **A** carboxylated sulphonylurea and pullulan and **B** dextran monosuccinyl metronidazole

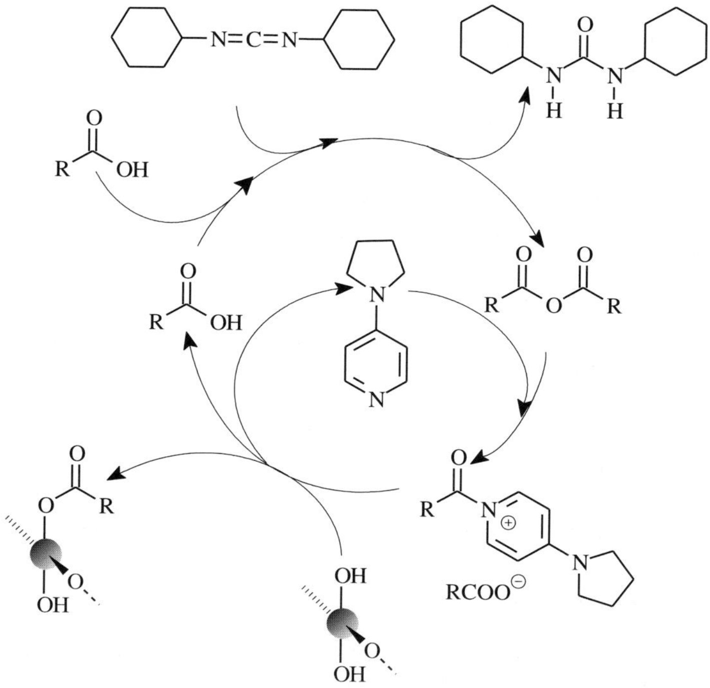

B)

Fig. 5.27. Scheme of the esterification of cellulose using PP/DCC (adapted from [223])

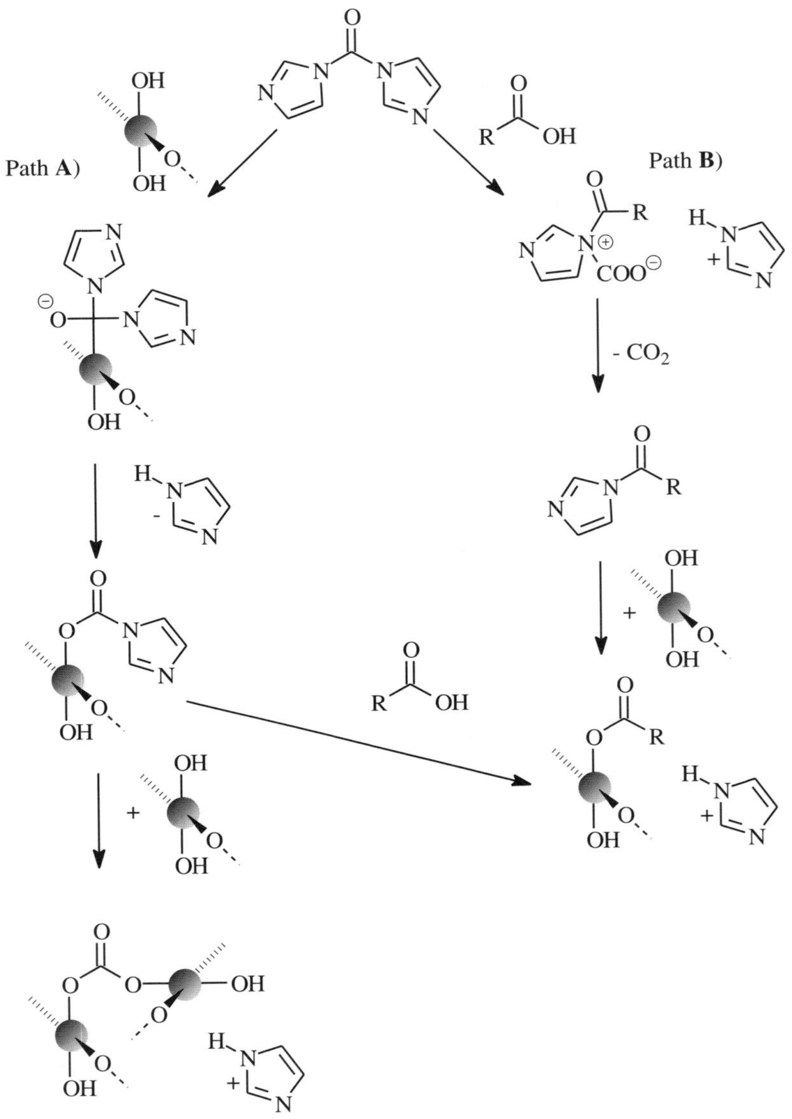

Fig. 5.28. Reaction paths leading exclusively to esterification (path **B**) or cross-linking (path **A**) if the polysaccharide is treated with CDI in the first step (adapted from [212])

in 1962 [228]. It is particularly suitable for the functionalisation of the biopolymers because, during the conversion, the reactive imidazolide of the acid is generated and the by-products formed are only CO_2 and imidazole (Fig. 5.28). The reagents and by-products are non-toxic. The imidazole is freely soluble in a broad variety of solvents including water, alcohol, ether, chloroform and Py, and can be easily removed. In addition, the pH is not strongly changed during the conversion, resulting in negligible chain degradation.

In comparison to DCC, the application of CDI is much more efficient, avoids most of the side reactions, and allows the use of DMSO (good solvent for most of the complex carboxylic acids) as solvent. In the case of CDI, no oxidation is observed and no decomposition of DMSO occurs (cf. no odour of dimethylsulphide).

The conversion is generally carried out as a one-pot reaction in two stages. To start with, the acid is transformed with the CDI to give the imidazolide. The conversion of the alcohol in this first step is also possible for the esterification but yields undesired cross-linking by carbonate formation in the case of a polyol (see Fig. 5.28). However, this process can be used for the defined cross-linking of starch [138].

The imidazolide of the carboxylic acid should always be firstly synthesised. Model reactions and NMR spectroscopy (Fig. 5.29) with acetic acid confirm that, during a treatment at room temperature, CDI is completely consumed within 6 h (compare Fig. 5.28). Thereby, the tendency of cross-linking of unreacted CDI leading to insoluble products is avoided.

Basic investigations towards conditions for coupling by the use of butyric acid and dextran confirm that the imidazolide is formed within 2 h and the reaction at RT for 17 h yields a butyrate containing 92% of the acid applied. Only 0.25% N is found in the ester. The solvent has a pronounced influence. In the case of dextran, the solvent of choice is the mixture formamide/DMF/CH_2Cl_2 [212]. PP is used as catalyst in this process.

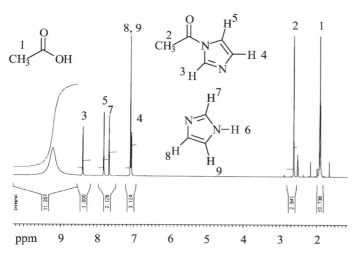

Fig. 5.29. ^{1}H NMR spectroscopic investigation of the in situ activation of acetic acid with CDI, confirming complete consumption of the CDI to the acetyl imidazolide

To ensure complete consumption of the CDI for all types of acids, the reaction towards the imidazolide is carried out most commonly for 16 h at 60 °C with equimolar ratios of acid/CDI. The intermediately formed imidazolide is unified with the polysaccharide solution to give the ester. The reaction is possible without catalysis or by applying PP, DMAP or alcoholates (e.g. potassium methanolate) in catalytic amounts. The utilisation of a catalyst may cause problems during the product isolation. Usually, precipitation and washing with ethanol is sufficient to obtain white, highly soluble, pure products.

Although CDI was applied as early as 1972 as reagent for the esterification of starch and dextran, it has only rarely been used up to now. Its renaissance during the last few years may be due to the fact that it became an affordable, commercially available product.

Among the first attempts for the esterification of polysaccharides via CDI is the introduction of amino acids into dextran. In addition to CDI, N,N'-(thiocarbonyl)diimidazole can be utilised to obtain the corresponding imidazolide [229]. The amino acids utilised are glycine, L-leucine, L-phenylalanine, L-histidine, and L-alanyl-L-histidine. They are protected with N-trifluoroacetyl-, N-benzyloxycarbonyl and 2,4-dinitrophenyl moieties. The protecting groups can be removed after the esterification of the polysaccharide by hydrolysis or hydrogenation over Pd catalyst [215].

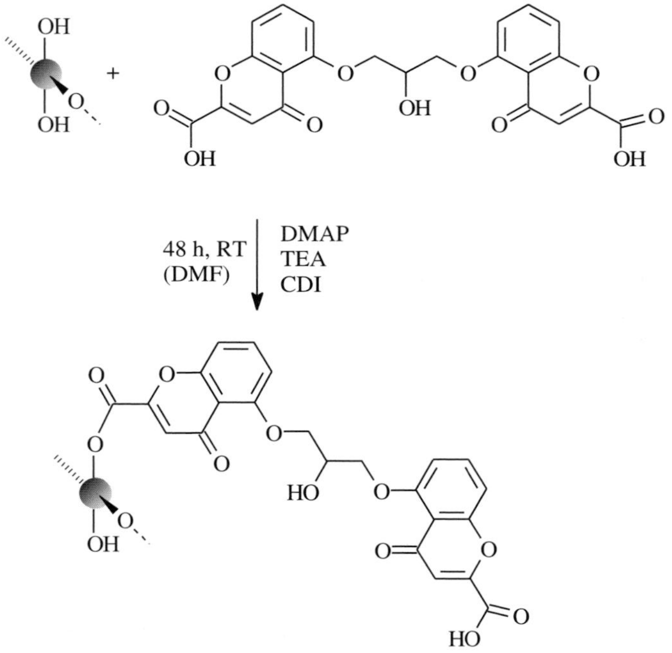

Fig. 5.30. Esterification of dextran with cromoglycic acid using in situ activation with CDI (adapted from [230])

The cromoglycic acid antiasthmatic drug (see Chap. 10) is covalently bound to dextran (Fig. 5.30). The acid is transferred into the imidazolide with CDI in DMF in the presence of TEA and DMAP within 5 h at RT. The conversion with dextran dissolved in DMF is achieved within 48 h at RT. The procedure gives high yields (up to 50%), with derivatives containing between 0.8 and 40% (w/w) of the acid (DS can not be calculated because there is no structural information, excluding the intermolecular esterification of the acid). Comparison with a route involving chlorination of the free acid in a first step, followed by reaction with dextran in formamide, results in low yields (1.5%) of an ester containing only 2.5% (w/w) cromoglycic acid [230].

CDI can also be utilised for the introduction of ester-containing substituents by coupling OH groups of the ester moiety with OH groups of the polysaccharide via a carbonate function. The route is shown in Fig. 5.31. By means of this method, a new class of polymerisable dextrans with hydrolysable groups becomes accessible [231].

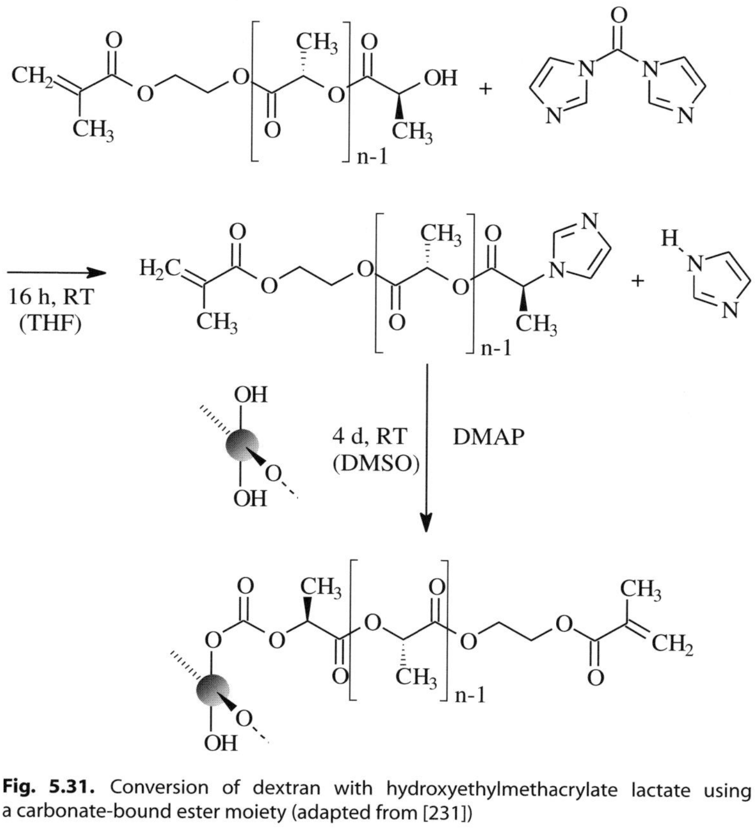

Fig. 5.31. Conversion of dextran with hydroxyethylmethacrylate lactate using CDI, yielding a carbonate-bound ester moiety (adapted from [231])

In the case of starch, the esterification applying CDI is used for the introduction of long-chain aliphatic esters. It is possible both in aqueous media and in suspension, e.g. halogenated hydrocarbons [138]. Introduction of aliphatic acids (acetate to stearate) and of di- and tricarboxylic acids is achieved but with low DS in the range of 0.01 to 0.15 (see Table 5.3).

Comparison of results of the esterification of starch with fatty acids (C_{14}–C_{18}) in DMAc/LiCl applying carboxylic acid chloride, in situ activation of the fatty acid with TosCl, and with CDI is given in Table 5.22. In situ activation with CDI leads to much less chain degradation and side reactions [168].

Table 5.22. Esterification of starch with long-chain aliphatic carboxylic acids using the corresponding chlorides and in situ activation with TosCl and CDI (adapted from [168])

Conditions								Product
Reagent	Molar ratio					Time	Temp.	DS
	AGU	Reagent	Coupling agent		Py	(h)	(°C)	
			Type	Equivalent				
Myristoyl chloride	1	4.5	–		5.4	6	100	2.69
Palmitoyl chloride	1	4.5	–		5.4	6	100	2.70
Stearoyl chloride	1	4.5	–		5.4	6	100	2.17
Myristic acid	1	6.0	TosCl	6	12.0	24	50	1.87
Palmitic acid	1	6.0	TosCl	6	12.0	24	50	1.18
Stearic acid	1	6.0	TosCl	6	12.0	24	50	2.17
Myristic acid	1	3.0	CDI	3	–	24	80	1.78
Palmitic acid	1	3.0	CDI	3	–	24	80	1.52
Stearic acid	1	3.0	CDI	3	–	24	80	1.65

It is noteworthy that esterification of starch with long-chain aliphatic acids using the imidazolide is also accomplished if the corresponding acid chlorides are converted with imidazole. The derivatisation is a homogeneous process in DMSO applying potassium methanolate as catalyst. The imidazole can be recovered. A summary of conditions and results for this path is given in Table 5.23 [232].

Table 5.23. Esterification of starch with long-chain aliphatic carboxylic acid imidazolides formed from the acid chloride and imidazole (adapted from [232])

Conditions					Product
Imidazolide	Molar ratio		Time (h)	Temp. (°C)	DS
	AGU	Reagent			
Octanoyl	1	2	3	90	1.55
Dodecanoyl	1	2	3	90	1.90
Hexadecanoyl	1	2	3	90	1.66

Table 5.24. DS values for starch cinnamates, furan-2-carboxylic acid esters, and 3-(2-furyl)-acrylic acid esters prepared homogeneously with CDI activation (molar ratio AGU:carboxylic acid:CDI = 1:3:3, adapted from [232])

Conditions			Product
Acid	Solvent	Temp. (°C)	DS
Cinnamic	DMAc/LiCl	50	0.89
Furan-2-carboxylic	DMSO	50	0.98
3-(2-Furyl)-acrylic	DMAc/LiCl	40	0.58
3-(2-Furyl)-acrylic	DMAc/LiCl	50	0.79
3-(2-Furyl)-acrylic	DMAc/LiCl	80	1.07

CDI can be used for the mild introduction of reactive unsaturated moieties into the starch backbone. Cinnamates, furan-2-carboxylic acid esters, and 3-(2-furyl)-acrylic acid esters of starch can be obtained in DMAc/LiCl and DMSO (Table 5.24, [233]).

Results of investigations concerning the potential of the in situ activation with CDI for a wide variety of carboxylic acids with chiral, (−)-menthyloxyacetic acid, unsaturated, 3-(2-furyl)-acrylcarboxylic acid, heterocyclic, furan-2-carboxylic acid, crown ether, 4′-carboxybenzo-18-crown-6, and cyclodextrin, carboxymethyl-β-cyclodextrin containing moieties are available by the conversion of cellulose, and will be discussed in detail (Fig. 5.32, [234]).

A reaction with (−)-menthyloxyacetic acid in situ activated with CDI can be carried out simply by mixing the solution of the imidazolide prepared in DMAc and the cellulose in DMAc/LiCl, and increasing the temperature to 60 °C. Pure (−)-menthyloxyacetic acid esters of cellulose with DS as high as 2.53 are obtained by precipitation in methanol and filtration (Table 5.25). The cellulose esters are characterised in terms of structure and DS by means of FTIR spectroscopy, elemental analysis, ^{1}H- and ^{13}C NMR spectroscopy, and additionally by ^{1}H NMR spectroscopy after peracylation. Cellulose (−)-menthyloxyacetate yields well-resolved NMR spectra.

Table 5.25. Esterification of cellulose with (−)-menthyloxyacetic acid via in situ activation with CDI (adapted from [234])

Reaction conditions			Product				
Molar ratio			DS	Solubility			
AGU	Acid	CDI		DMSO	DMF	Acetone	CHCl$_3$
1	2.5	2.5	0.20	+	+	−	−
1	5.0	5.0	1.66	−	+	+	+
1	7.5	7.5	2.53	−	+	+	+

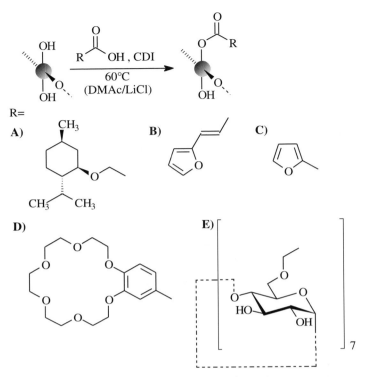

Fig. 5.32. Conversion of cellulose with carboxylic acids applying in situ activation with CDI yielding the esters of **A** (−)-menthyloxyacetic acid; **B** 3-(2-furyl)-acrylcarboxylic acid; **C** furan-2-carboxylic acid; **D** 4′-carboxybenzo-18-crown-6; **E** carboxymethyl-β-cyclodextrin (adapted from [234])

Figure 5.33 shows the ^{13}C NMR spectrum of cellulose (−)-menthyloxyacetate (DS 1.66) recorded in CDCl$_3$. The highly functionalised product is even soluble in easily evaporable solvents including THF and chloroform, which is desired for the transformation into membranes or beads.

Signal assignment was achieved by comparison with simulated spectra and by measuring DEPT135 NMR spectra (Fig. 5.33). Besides signals for the carbons of the AGU (103.7–60.1 ppm), resonances assigned to the carbon atoms of the menthyloxyacetate moieties are visible between 81.0 (C-9) and 16.9 ppm (C-16) (for complete assignment, see Chap. 12). The carbon atoms C-16 and C-17 are not chemically equivalent because of the chirality of the substituent, and give two separate signals at 16.9 and 21.3 ppm.

The peak for C-6, influenced by esterification in O-6, appears at 62.4 ppm (C-6$_s$). In addition, the spectrum shows a signal at 101.6 ppm, corresponding to C-1 adjacent to the C-2 atom bearing a menthyloxyacetate unit. Comparison of the intensities of signals related to substituted and unsubstituted C-2 and C-6 reveals that substitution with the bulky menthyloxyacetate moiety proceeds faster at the primary OH.

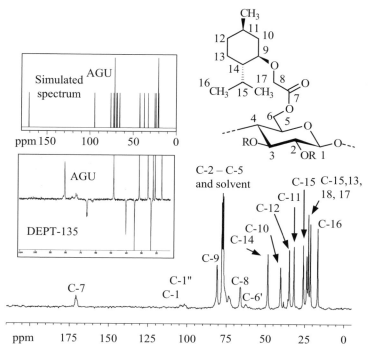

Fig. 5.33. ^{13}C NMR spectrum of cellulose (–)-menthyloxyacetate (DS 1.66, in CDCl$_3$), compared with its simulated spectrum and a DEPT135 NMR spectrum (adapted from [234])

For complete assignment of the ^{1}H NMR spectrum of this complex polysaccharide ester, simulation and a variety of two-dimensional techniques including ^{1}H,^{1}H-COSY-DQF-, HSQC-DEPT- and HSQC-TOCSY-NMR spectra (Fig. 5.34) are necessary.

The protons of the menthyloxyacetate moiety are visible in the range from 0.82 to 3.19 and at 4.13 ppm (for complete assignment, see experimental section of this book). The two sets of protons of the methyl moieties of C-16 and C-17 have different chemical environments and therefore show two separate signals at 0.82 and 0.92 ppm. The presence of chiral carbon atoms results in splitting of the signals of the protons in position 10, 12 and 13. Therefore, peaks at 0.87 and 1.67 (H-12 and H-12*), at 0.93 and 2.08 (H-10 and H-10*), and at 0.99 and 1.64 ppm (H-13 and H-13*) can be found. Signals of the AGU are observed at 3.42–4.98 ppm. The NMR spectroscopy confirms the structural homogeneity of the ester. There are no hints for side reactions or impurities.

The conversion of cellulose with camphor-10-sulphonic acid via in situ activation with CDI can not be used to obtain a chiral sulphonic acid ester of cellulose. Only very small amounts of sulphonic acid ester functions can be introduced, in agreement with results of the chemistry of low-molecular mass alcohols showing a much lower efficiency of CDI for the preparation of sulphonic acid esters [228].

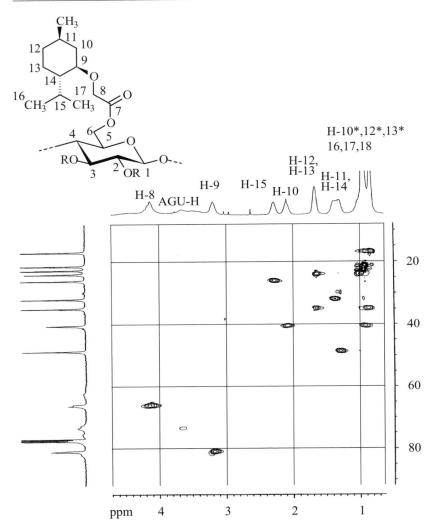

Fig. 5.34. HSQC NMR spectrum of cellulose (−)-menthyloxyacetate (DS 1.66, in CDCl₃). R = H or the ester moiety, according to the DS and the distribution of the functional groups (adapted from [234])

In the case of unsaturated esters, the double bonds can be exploited for subsequent cross-linking or for grafting reactions. Nevertheless, these reactions need to be suppressed during the esterification with the polysaccharide to obtain soluble products. Therefore, mild reaction conditions are indispensable for reactive unsaturated acids. In this regard, CDI is very helpful, although the preparation of esters with terminal double bonds may be combined with the introduction of covalently bound imidazole units, which is shown for the conversion with acrylic acid in Fig. 5.35.

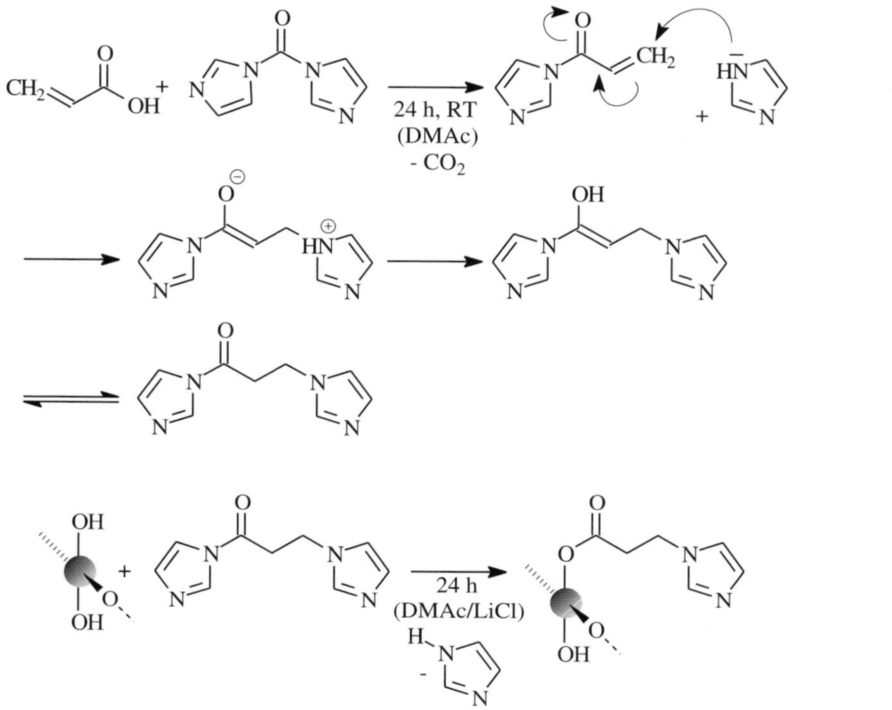

Fig. 5.35. Mechanistic considerations for the binding of imidazole containing esters via reaction of polysaccharides with acrylic acid (adapted from [233])

Nonetheless, via in situ activation with CDI, the preparation of 3-(2-furyl)-acrylcarboxylic acid esters of cellulose is possible. A maximum DS of 1.52 is obtained, and the sample is soluble in the freshly precipitated form. Isolation and drying gives an insoluble product, which is obviously due to a spontaneous cross-linking process. Structural analysis by means of ^{13}C NMR spectroscopy in DMSO-d_6 can be carried out for derivatives with lower DS showing characteristic signals, i.e. at 165.3 ppm for the carbonyl carbon atom of the ester, at 150.1 (C-10), 145.8 (C-13) 116.5 (C-12) and 112.5 ppm (C-11) for the furan ring, at 145.8, (C-9), 131.6 ppm (C-8) for the double bond, and from 60.2 to 102.9 ppm for the AGU. A preferred functionalisation of position 6 and no impurities (oxidation reactions, imidazole) are found. DS values calculated from ^1H NMR spectra of completely functionalised samples by subsequent perpropionylation of the remaining OH groups in CDCl$_3$ are summarised in Table 5.26. Both the pure 3-(2-furyl)-acrylcarboxylic acid esters of cellulose and the propionylated samples need to be stored in the dark. Otherwise, they become insoluble due to cross-linking.

Preparation of a furan-2-carboxylic acid ester of cellulose is achieved using CDI for the in situ activation. DS values up to 1.97 are obtainable (Table 5.26, entry **6**). Structural evidence is gained by ^{13}C NMR spectroscopy (Fig. 5.36). Signals at

Table 5.26. Esterification of cellulose with furan-2-carboxylic acid, 3-(2-furyl)-acrylcarboxylic acid, and 4′-carboxybenzo-18-crown-6 via in situ activation with CDI (adapted from [234])

Conditions					Product		
Entry	Carboxylic acid	Molar ratio			DS	Solubility	
		AGU	Acid	CDI		DMSO	DMF
1	3-(2-Furyl)-acryl-	1	2.5	2.5	0.52	+	–
2		1	5.0	5.0	1.14	+	+
3		1	7.5	7.5	1.52	+	–
4	Furan-2-	1	2.5	2.5	0.80	+	+
5		1	5.0	5.0	1.49	+	+
6		1	7.5	7.5	1.97	+	+
7	(Benzo-18-crown-6)-4′-	1	2.3	2.3	0.40	+	–

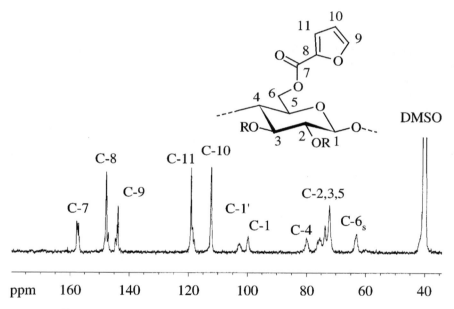

Fig. 5.36. ^{13}C NMR spectrum of a furan-2-carboxylic acid ester of cellulose (DS = 1.91, adapted from [234])

112.1, 118.8, 143.4 and 157.4 ppm show that the unsaturated system is stable under the reaction conditions applied. From the peak at 63.1 ppm and the occurrence of two signals at 100.2 (C-1′) and 102.4 ppm (C-1), a complete functionalisation at the primary OH-group and partial functionalisation in position 2 can be concluded for samples with DS values higher than 1.4 (Table 5.26, entries **4–6**). These findings are comparable with results for furan-2-carboxylic acid esters synthesised homogeneously by conversion of cellulose in DMAc/LiCl applying the acid chloride [170, 235].

Table 5.27. GPC analysis (in DMSO) of 3-(2-furyl)-acrylcarboxylic acid ester of cellulose (Table 5.26, entry **3**) and furan-2-carboxylic acid esters (entries **4, 6**)

Entry	DS	M_w (g mol^{-1})	DP
3	1.52	7.21×10^4	208
4	0.80	6.55×10^4	275
6	1.97	6.91×10^4	200

GPC studies for the unsaturated cellulose esters reveal a bimodal distribution, as usually observed for partially functionalised cellulose derivatives. The low-molecular mass fraction can be assigned to polymers dissolved in a molecular-dispersed manner. The depolymerisation during the conversion is rather small (Table 5.27). Product **4** possesses a DP of 275 (the starting cellulose Avicel® has a DP of 280).

For the cellulose furan-2-carboxylic acid esters, the cross-linking process, which can be exploited for subsequent modification (see Application) of the derivative, e.g. in membrane shape, can be initiated by means of UV irradiation [234].

The introduction of crown ether moieties into the cellulose backbone is usually achieved by reactive coupling of amino-functionalised crown ethers with cyanuric chloride onto cellulose diacetate [236]. The material is valued as basis for an alkaline-ion sensitive electrode. A more efficient approach is the homogeneous esterification of cellulose with 4'-carboxybenzo-18-crown-6. Homogeneous conversion in DMAc/LiCl and in situ activation of the carboxylic acid with CDI is the reaction of choice. Cellulose 4'-carboxybenzo-18-crown-6 esters can be obtained

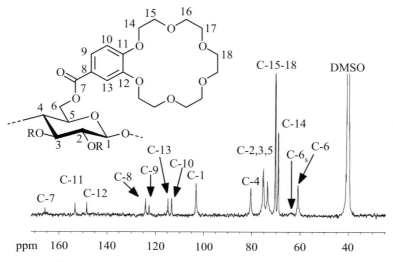

Fig. 5.37. ^{13}C NMR spectrum of a cellulose 4'-carboxybenzo-18-crown-6 ester (DS = 0.4, adapted from [234])

as white substances that dissolve in DMSO (Table 5.26). The polymer yields well-resolved ^{13}C NMR spectra, as shown in Fig. 5.37. In addition to the signals for the carbons of the modified AGU (103.2 to 60.1 ppm), resonances assigned to the carbon atoms of the carbonyl group of the ester function at 165.8 ppm, resonances of the crown ether moiety at 69.3, 70.4 ppm, and those for the carbons of the aromatic system at 113.3, 114.8, 148.5 and 153.3 ppm are observed.

The ^{1}H,^{1}H-COSY NMR spectrum (CDCl$_3$) of perpropionylated cellulose 4′-carboxybenzo-18-crown-6 ester is shown in Fig. 5.38. These complex esters give only seven signals for the protons of the AGU, meaning that the pattern of substitution does not yield signal splitting. ^{1}H NMR spectroscopy can still be applied

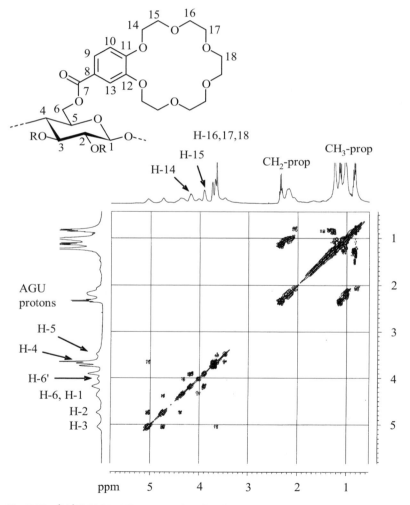

Fig. 5.38. ^{1}H,^{1}H-COSY NMR spectrum (CDCl$_3$) of a perpropionylated cellulose 4′-carboxybenzo-18-crown-6 ester (adapted from [324])

for the DS determination (see Sect. 8.3). The DS is calculated from the ratio of the spectral integrals of the H-3 and H-2 protons of the repeating unit at 5.00 (H-3), 4.85 ppm (H-2) versus the CH_2-protons of the propionate moiety at 2.04–2.39 ppm and, gives a value of 0.40.

Cyclodextrin moieties may be bound to the polysaccharide backbones via formation of a Schiff's base [237] of chitosan with 2-O-(formylmethyl)-β-cyclodextrin or via immobilisation of cyclodextrin on polysaccharides using cross-linking agents, e.g. polymeric anionic reactive compounds [238]. An efficient alternative is the conversion of cellulose with carboxymethyl-β-cyclodextrin. The commercially available sodium salt form of the cyclodextrin derivative is converted to the free acid form by treatment with methanolic HCl (20% w/w). The in situ activation with CDI and the reaction with the polysaccharide should be carried out in one step in DMSO. Otherwise, i.e. during a separate activation step, the carboxymethyl-β-cyclodextrin reacts according to an intermolecular cross-linking of the imidazolide formed with remaining OH groups, and yields an insoluble precipitate. Thus, cellulose dissolved in DMAc/LiCl is treated directly with carboxymethyl-β-cyclodextrin and CDI for 16 h at 80 °C, leading to an insoluble product. The formation of the ester can be confirmed by the carbonyl group signal in the FTIR spectrum (ν(C=O) at 1724 cm^{-1}). Moreover, signals of unesterified carboxy moieties of the cyclodextrine not involved in ester formation are observed at 1655 and 1426 cm^{-1}.

The introduction of bulky alicyclic functions is performed by conversion of cellulose with adamantane carboxylic acid, yielding DS values up to 1.41 [169]. Comparison with reactions using the acid chloride and in situ activation with TosCl gives similar results to those of the esterification with fatty acids, i.e. slightly lower efficiency using CDI but less side reactions (white products) and easier workup. A ^{13}C NMR spectrum of adamantoyl cellulose (DS 0.68, Fig. 5.39) shows

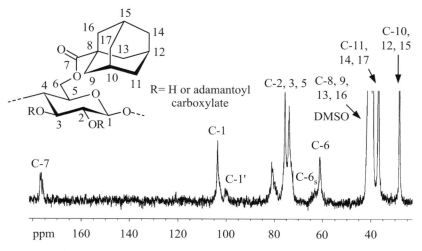

Fig. 5.39. ^{13}C NMR spectrum of a cellulose adamantane-1-carboxylic acid ester (DS = 0.68, adapted from [169])

signals for the carbons of the modified AGU (103.7 to 60.1 ppm) and resonances of the carbon atoms of the adamantoyl ester moieties at 28.2 (C-10, C-12, C-15), 36.8 (C-11, C-14, C-17) and 39.16 ppm (C-9, C-13, C-16). The C-8 signal is overlapped by the solvent. The splitting of both the C-6 (63.1 ppm, C-6$_s$) and the C-1 signal (100.4 ppm for C-1 adjacent to a C-2 atom bearing an adamantoyl moiety) shows a roughly even distribution of substituents over the AGU. Despite the steric bulk of the adamantoyl moiety, no pronounced regioselectivity is observed.

In addition to DMAc/LiCl, DMSO/TBAF is an appropriate reaction medium for homogeneous acylation of cellulose applying in situ activation with CDI. Results of reactions of cellulose with acetic-, stearic-, adamantane-1-carboxylic- and furan-2-carboxylic acid imidazolides are summarised in Table 5.28.

Table 5.28. Homogeneous acylation of cellulose dissolved in DMSO/TBAF with different carboxylic acids, mediated by CDI

Entry	Conditions				Product	
	Carboxylic acid	Molar ratio			DS	Solubility
		AGU	Acid	CDI		
1	Acetic	1	3	3	0.51	DMSO, DMAc
2	Stearic	1	2	2	0.47	DMSO
3	Stearic	1	3	3	1.35	DMSO
4	Adamantane-1-carboxylic	1	2	2	0.50	DMAc/LiCl
5	Adamantane-1-carboxylic	1	3	3	0.68	DMSO, DMAc
6	Furan-2-carboxylic	1	3	3	1.91	DMSO, DMAc, Py

NMR spectra confirm that pure cellulose esters are obtained by precipitation in ethanol and no side reactions occur (tetra-N-alkylammonium fluorides typically decompose under anhydrous conditions [174]). Washing with ethanol is sufficient to completely remove the imidazole, as can be concluded from the lack of signals at 7.13 and 7.70 ppm (^{1}H NMR data).

The conversion of the dissolved cellulose with furan-2-carboxylic acid imida-zolide (with a stoichiometry of AGU:reagent of 1:3) yields a rather high DS of 1.91 (Table 5.28, entry **6**), which corresponds to a remarkable reaction efficiency of 63%. NMR spectroscopy reveals the same pattern of substitution as that determined for a cellulose furan-2-carboxylic acid ester prepared in DMAc/LiCl. The aliphatic es-ters (cellulose acetate and stearate; entries **1–3**) show DS values up to 1.35. Reaction of cellulose in DMAc/LiCl using the carboxylic acid anhydrides leads to DS values of 1.2 for the acetate and of 2.1 for the stearate under similar conditions. This indi-cates a comparably high reactivity of the imidazolides of shorter carboxylic acids (C_2–C_4) towards hydrolysis caused by the water in the reaction medium (TBAF trihydrate is used). Imidazolides of long-chain aliphatic acids are less reactive in this solvent.

Cellulose adamantane-1-carboxylic acid esters obtained in DMSO/TBAF exhibit DS values of up to 0.68. The amazing conclusion from ^{13}C NMR spectroscopical studies (Fig. 5.39) is that the functionalisation with the bulky adamantoyl unit occurs more pronounced at position 2 if DMSO/TBAF is applied as medium. The reason might be partial hydrolysis of the ester formed during the reaction. GPC studies show only small depolymerisation (approximately 13%).

One can conclude that homogeneous esterification of cellulose with carboxylic acid/CDI in DMAc/LiCl and DMSO/TBAF with in situ-prepared carboxylic acid imidazolides is one of the simplest and most widely usable synthesis pathways for the preparation of a very broad variety of pure cellulose esters, which can easily be extended to other polysaccharides. In contrast to DCC or TosCl as reagents for in situ activation, the CDI is associated with no significant side reactions, even when DMSO is used as solvent, if the CDI is completely transformed to the imidazolide in the first step. The products obtained are only slightly degraded, pure, and highly soluble compounds. In the case of the reaction of carboxylic acids with active protons, e.g. OH, NH$_2$, terminal double- or triple bonds, protection prior to the esterification is necessary.

The combination DMSO/TBAF as solvent and CDI as reagent for in situ activation is one of the most convenient homogeneous paths for cellulose esterification, even for inexperienced personal. In the case of aromatic acids, the path is superior to the conversion in DMAc/LiCl in terms of efficiency and simplicity. Although the yields are diminished by the presence of water in the case of aliphatic acid imidazolides, the procedure is one of the most promising tools for the synthesis of cellulose derivatives with complex ester moieties, e.g. unsaturated and chiral moieties not accessible via the carboxylic acid anhydrides and -chlorides.

5.2.4 Iminium Chlorides

A mild and efficient method is the in situ activation of carboxylic acids via iminium chlorides. They are simply formed by conversion of DMF with a variety of chlorinating agents, including phosphoryl chloride, phosphorus trichloride and, most frequently, oxalyl chloride and subsequent reaction with the acid. During the reaction of acid iminium chlorides with alcohols, mostly gaseous side products are formed and the solvent is regenerated (Fig. 5.40, [239]).

The reaction is very mild. The synthesis of the intermediate is carried out at $-20\,^\circ$C. The complex formed is stable and no side reactions, such as the formation of HCl or the acid chloride, are observed. Consequently, it is a suitable process for polysaccharide esterification.

Acylation of cellulose with the long-chain aliphatic acids (stearic acid and palmitic acid), the aromatic acid 4-nitrobenzoic acid and adamantane-1-carboxylic acid is easily achieved. The formation of the iminium chloride and the conversion with the carboxylic acid are carried out as "one-pot reaction", i.e. DMF is cooled to $-20\,^\circ$C, oxalyl chloride is added very carefully and, after the gas formation ceases, the carboxylic acid is added. NMR spectroscopy reveals that the conversion succeeds with measurable yield in the case of acetic acid. The mixture is added to

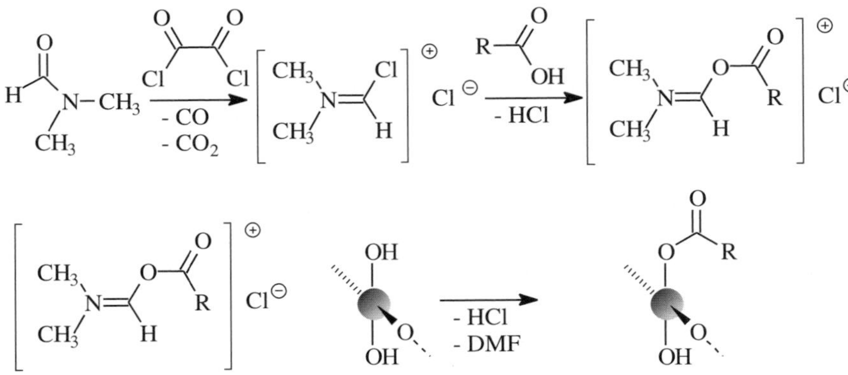

Fig. 5.40. Preparation of cellulose esters via in situ activation by iminium chlorides (adapted from [239])

a solution of cellulose in DMAc/LiCl and treated at 60 °C for 16 h. The purification is simple practically because most of the gaseous by-products are liberated from the reaction mixture and, during the last step, DMF is formed (Fig. 5.40).

In the case of fatty acids, the cellulose ester floats on the reaction mixture if stirring is stopped at the end of the conversion, and can be isolated in very good yields simply by filtration and washing with ethanol. A summary of reaction conditions and results is given in Table 5.29.

Table 5.29. Esterification of cellulose dissolved in DMAc/LiCl via the iminium chlorides of different carboxylic acids (adapted from [239])

Entry	Carboxylic acid	Molar ratio			Product	
		AGU	Acid	Oxalyl chloride	DS	Solubility
1	Stearic	1	3	3	0.63	DMSO/LiCl
2	Stearic	1	5	5	1.84	THF, CHCl₃
3	Palmitic	1	6	6	1.89	DMSO, DMAc, THF
4	Adamantane-1-carboxylic	1	1	1	0.47	DMSO/LiCl
5	Adamantane-1-carboxylic	1	3	3	1.20	DMAc, DMSO, DMF
6	Adamantane-1-carboxylic	1	6	6	0.66	DMSO
7	4-Nitrobenzoic	1	1	1	0.30	DMSO/LiCl
8	4-Nitrobenzoic	1	2	2	0.52	DMSO
9	4-Nitrobenzoic	1	3	3	0.94	DMSO
10	4-Nitrobenzoic	1	6	6	0.66	DMSO/LiCl

The procedure is suitable for the synthesis of all types of cellulose esters. It is particularly efficient for the esterification with aliphatic and alicyclic carboxylic acids (Table 5.29, entries 1–6). DS values as high as 1.89 can be achieved, yielding polymers soluble in THF. Thus, in terms of efficiency, results comparable to the conversion with carboxylic acid chlorides or via activation with TosCl are obtained. Increasing DS values are observed for molar ratios of carboxylic acid to AGU of up to 5:1. If the ratio is about 6:1, the solutions become highly viscous or even gel-like during the reaction, resulting in decreasing DS (Table 5.29, entries **3** and **10**). Acetylation of cellulose yields highly functionalised esters (analysed by FTIR) that are insoluble in common organic solvents. This insolubility is also observed for cellulose acetates prepared with acetyl chloride (without base). It is unknown if this behaviour is due to an unconventional superstructure, e.g. caused by the complete acetylation of the primary hydroxyl function and/or an uneven distribution of the acetyl groups within the polymer chains.

A representative well-resolved ^{1}H NMR spectrum (CDCl$_3$) of cellulose 4-nitrobenzoate (Table 5.29, entry **9**) after perpropionylation is shown in Fig. 5.41, as example for an ester synthesised via the iminium chloride activation. The spectrum contains signals for the AGU at 3.46–5.04 ppm (H-1–H-6), for the aromatic protons of the nitrobenzoate moiety at 7.79–8.31 ppm (H-7, 8), and for the propionate at 2.10 ppm (H-9) and 0.99 ppm (H-10). No side reactions, e.g. chlorination or oxidation, are observed. This is confirmed by ^{13}C NMR- and FTIR spectroscopy. Nevertheless, the esters synthesised contain up to 2% chlorine.

GPC experiments corroborate the mildness of the conversion, resulting in negligible degradation of the cellulose backbone. DP values of 240 for cellulose adamantate (DS 1.20, Table 5.29, entry **5**), 280 for cellulose 4-nitrobenzoate (DS 0.52, Table 5.29, entry **8**) and 250 for cellulose stearate (DS 1.84, Table 5.29, entry **2**) are obtained if Avicel® with DP 280 is applied as starting cellulose. Thus, esterifica-

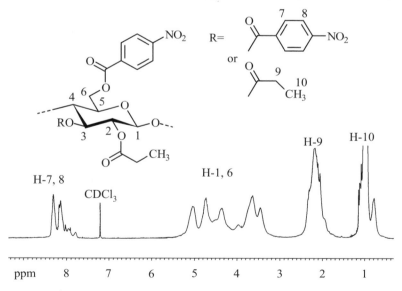

Fig. 5.41. ^{1}H NMR spectrum of a perpropionylated cellulose 4-nitrobenzoate (DS = 0.94, adapted from [239])

tion via iminium chlorides is much milder than conversion via in situ activation with TosCl or functionalisation with the acid chlorides. Consequently, this type of esterification combines a high efficiency with very mild reaction conditions.

It might be possible to exploit this path for the synthesis of sophisticated or sensitive esters, e.g. with unsaturated or chiral moieties. It is the most inexpensive in situ activation established to date, which might be applied at a large scale. It is suitable for the homogeneous esterification of polysaccharides because it uses DMF as solvent and reagent, which is an appropriate medium for polysaccharide functionalisation. This method could significantly broaden the scope of this very promising process.

5.3 Miscellaneous New Ways for Polysaccharide Esterification

5.3.1 Transesterification

For the preparation of long-chain aliphatic esters of polysaccharides, acylation reactions of starch and cellulose via transesterification with methyl esters of palmitic- and stearic acid have been studied [240]. The heterogeneous conversion of cellulose suspended in DMF with methyl stearate yields a derivative with DS values up to 0.38, implying the reaction is comparably inefficient [241]. To increase the efficiency of this method, different reagents and catalysts are applied. Potassium methoxide significantly increases the DS of starch palmitates prepared with methyl palmitate in DMSO [242]. The concentration of the catalyst has a strong influence (Table 5.30). The maximum DS of 1.5 is achieved if a concentration of 0.1 mol

Table 5.30. Influence of the catalyst (potassium methoxide) concentration on the DS during the reaction of starch with methyl palmitate (adapted from [241])

Conditions				Product
Molar ratio			Temp. (°C)	DS
AGU	Methyl palmitate	KOCH$_3$		
1	1	0.100	100	0.59
1	3	0.010	100	0.18
1	3	0.025	100	0.60
1	3	0.050	100	0.86
1	3	0.100	80	0.71
1	3	0.100	90	0.90
1	3	0.100	100	1.10
1	3	0.100	110	0.86
1	3	0.200	100	0.98
1	5	0.100	100	1.48
1	10	0.100	100	1.52

catalyst per mol AGU is adjusted. Replacement of methyl palmitate with methyl esters of shorter-chain acids does not appear to affect the DS.

In new approaches for esterification of polysaccharides via transesterification, the vinyl esters of the carboxylic acids are predominantly exploited. During the conversion, the instable vinyl alcohol is formed and is immediately transformed into acetaldehyde, shifting the equilibrium towards the product side (Fig. 5.42). Among the catalysts for the transesterification of polysaccharides with vinyl esters are a variety of enzymes. An interesting development is the application of enzymes in anhydrous organic media. It has been assumed that enzymes are inactive in solvents such as DMSO or DMF caused deleterious changes in the secondary and tertiary structure.

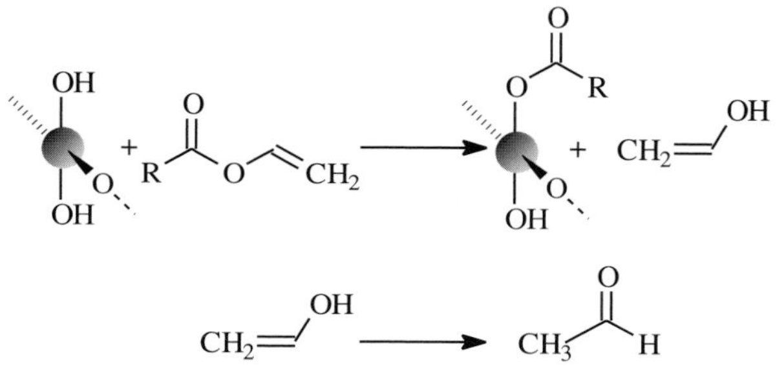

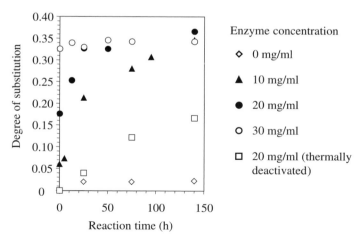

Fig. 5.42. Formation of acetaldehyde from vinyl alcohol formed during transesterification of polysaccharides with vinyl esters of carboxylic acids

Recent work shows that enzymes, e.g. Proleather and lipase, are partially soluble in organic media but remain active under these conditions [243]. Thus, the *Bacillus subtilis* protease Proleather FG-F has been used to catalyse the transesterification of inulin with vinyl acrylate in DMF [244]. The DS can be controlled by varying the molar ratio of vinyl acrylate to inulin and by the enzyme concentration (Fig. 5.43).

Enzyme concentration

◇ 0 mg/ml

▲ 10 mg/ml

● 20 mg/ml

○ 30 mg/ml

□ 20 mg/ml (thermally deactivated)

Fig. 5.43. Conversion of inulin with vinyl acrylate in the presence of different concentrations of the enzyme Proleather FG-F (adapted from [244])

The maximum DS is 0.45. Structure analysis by means of ^{1}H, ^{13}C, ^{1}H,^{1}H-COSY and HBMC NMR spectroscopy reveals preferred functionalisation at position 6 of the fructofuranoside residue.

Dextran can be acylated with vinyl acrylate in the presence of Proleather FG-F and lipase AY, a protease and lipase from *Bacillus* sp. and *Candida rugosa* in anhydrous DMSO. Structure analysis by NMR spectroscopy indicates functionalisation of positions 2 and 3 of the AGU in equal amounts [243]. The efficiency of the reaction and the DS accessible in the presence of Proleather FG-F is shown in Table 5.31.

Table 5.31. Dependence of the DS on the amount of reagent (calculated as theoretical DS) applied during the conversion of dextran with vinyl acrylate in the presence of the enzyme Proleather FG-F (adapted from [243])

DS		Efficiency (%)
Theoretical	Obtained	
0.10	0.072	71.4
0.20	0.151	75.7
0.30	0.224	74.6
0.40	0.315	78.9
0.50	0.370	74.1

The derivatisation of starch with carboxylic acid vinyl esters with Proteinase N is very efficient and highly regioselective in position 2, as discussed in Chap. 9 [245]. The transesterification needs to be carried out in anhydrous DMSO at 39 °C to suppress pure chemical (non-enzyme catalysed) esterification, which does not lead to selectively functionalised products. Results of optimisation experiments are summarised in Table 5.32, showing that the DS can be adjusted in a defined manner.

The enzymatically catalysed transesterification is used for selective functionalisation of starch nanoparticles, applying *Candida antarctica* Lipase B in its immobilised (Novozym 435) and free (SP-525) forms. The nanoparticles are converted in surfactant/isooctane/water microemulsions with vinyl stearate for 48 h at 40 °C to give starch esters with DS values up to 0.8. In contrast to the transesterification in DMSO, the reaction occurs regioselectively at position 6 of the RU. Even though *C. antarctica* Lipase B is immobilised within a macroporous resin, *C. antarctica* Lipase B is sufficiently accessible to the starch nanoparticles.

Candida antarctica Lipase B immobilised in inverse micelles with starch is also active and catalyses the acylation with vinyl stearate (24 h, 40 °C) to give a product with DS 0.5. After removal of the surfactant from the modified starch nanoparticles, they can be dispersed in DMSO or water, with retention of their nanodimensions [246]. Cellulose is acetylated by transesterification with vinyl

Table 5.32. Influence of the amount of reagent and of the time on the DS values for the transesterification of starch with vinyl acetate in the presence of Proteinase N (adapted from [245])

Conditions Molar ratio AGU	Vinyl acetate	Time (h)	Product DS
1	2.3	2	0.1
1	2.3	5	0.3
1	0.5	70	0.3
1	2.3	10	0.5
1	1.0	70	0.5
1	1.5	70	0.7
1	2.3	20	0.8
1	2.3	30	0.9
1	2.3	70	1.0
1	4.0	70	1.1

acetate under homogeneous reaction conditions in NMMO. However, the product obtained possesses a rather small DS of 0.3 [125].

In addition to enzyme-catalysed reactions with carboxylic acid vinyl esters, different salts are able to catalyse the conversion. The reaction in an organic medium such as DMSO catalysed by an inorganic salt gives comparable DS values to that of the Proteinase N (*Bacillus substilis*) [245]. In both cases, a remarkable selectivity of the reaction is observed (see Chap. 9). A summary of salts usable for the acylation procedure and the DS values obtained in the case of starch with vinyl acetate as well as the influence of the kind of starch used on the DS is given in Tables 5.33 and 5.34. The highest DS is achieved with K_2CO_3 as catalyst but the most valuable catalysts, in terms of regioselectivity, are Na_2HPO_4 and Na_2CO_3 and even NaCl shows a catalytic activity. However, the detailed mechanisms are unknown.

Table 5.33. DS values in function of the type of salt applied as catalyst during acetylation of starch with vinyl acetate (2.3 mol/mol AGU, 70 h, 40 °C, adapted from [245])

Reaction conditions Catalyst	Acylating agent	Starch acetate DS
DMAP, Py	Acetic anhydride	1.88
Na_2HPO_4	Acetic anhydride	1.00
Na_2HPO_4	Vinyl acetate	1.00
NH_4Cl	Vinyl acetate	0.95
NaCl	Vinyl acetate	1.00
K_2CO_3	Vinyl acetate	2.18
Na_2CO_3	Vinyl acetate	1.82

Table 5.34. DS values in function of the kind of polysaccharide by acetylation with vinyl acetate/Na_2HPO_4 (40 h, 70 °C, 2.3 mol vinyl acetate per mol AGU, 2% w/w Na_2HPO_4, adapted from [245])

Oligo/polysaccharide Type	Product DS
Hylon VII	1.00
Corn starch	0.92
Wheat starch	0.94
Potato starch	0.88
Waxy corn starch (Amioca powder)	0.82
Glycogen	0.10
α-Cyclodextrin	1.00
β-Cyclodextrin	1.00
Pullulan	0.75
Nigeran	1.00

In addition to simple acetylation, the procedure can be applied for the efficient preparation of a wide variety of different starch esters including long-chain aliphatic derivatives, halogen substituted derivatives and unsaturated esters, leading to products with pronounced selectivity to position 2, as found for the starch acetates. A summary of these esters is shown in Table 5.35.

This salt-catalysed transesterification can be used for the esterification of cellulose as well. An efficient alternative for the preparation of acetylated cellulose is the transesterification using the solvent DMSO/TBAF [27]. With vinyl acetate

Table 5.35. Starch esters obtainable via transesterification of the biopolymer with vinyl esters of carboxylic acids and different catalysts (adapted from [245])

Reaction conditions Vinyl-	Catalyst	Starch ester DS	Yield (%)	$\nu_{C=O}$ (cm^{-1})
Acetate	Na_2HPO_4	1.00	90	1730
Propionate	Na_2HPO_4	1.00	85	1728
Butanoate	Na_2HPO_4	1.00	70	1728
Laurate	Na_2HPO_4	0.70	78	1730
Chloroacetate	Na_2CO_3	1.00	82	1720
Pivalate	Na_2CO_3	1.10	90	1725
Benzoate	Na_2HPO_4	0.92	90	1718
Acrylate	Na_2HPO_4	0.90	80	1715
Methacrylate	Na_2HPO_4	0.92	88	1705
Crotonate	Na_2CO_3	0.95	82	1710
Cinnamate	Na_2CO_3	1.02	90	1703

Table 5.36. Acetylation of cellulose (2.9%) dissolved in DMSO/TBAF (16.6%) with vinyl acetate in the presence of a catalyst (mixture of KH_2PO_4 and Na_2HPO_4) at 40 °C for 70 h (adapted from [27])

Conditions Molar ratio		Catalyst (mg)	Product Partial DS		DS	Solubility
AGU	Vinyl acetate		O-6	O-2/3		
1	2.3	–	0.49	0.55	1.04	DMSO
1	2.3	20	0.52	0.55	1.07	Insoluble
1	1.5	20	0.39	0.24	0.63	Insoluble
1	10	20	0.98	1.74	2.72	DMSO

as acylating reagent, DS values up to 2.72 are obtainable. A summary of reaction conditions and results is given in Table 5.36.

The transesterification in DMSO/TBAF is much more efficient than acetylation with acetic anhydride. In the case of acetic anhydride, the lower DS is caused by the comparably fast hydrolysis of the reagent due to the water content of the solvent. In addition to the acetates, a variety of fatty acid esters and of aromatic acid esters can be very efficiently obtained, as shown in Table 5.37.

Table 5.37. Cellulose acylation in DMSO/TBAF (16.6%) with vinyl carboxylic acid esters (adapted from [27, 129])

Conditions Reagent	Molar ratio		Time (h)	Temp. (°C)	Cellulose ester DS
	AGU	Reagent			
Vinyl butyrate	1	2.3	70	40	0.86
Vinyl laurate	1	2.3	3	60	1.24 (sisal)
Vinyl laurate	1	2.3	70	40	1.47
Vinyl laurate	1	10	70	40	2.60
Vinyl benzoate	1	2.3	70	40	0.95

An interesting new catalyst is $[(Bu_2SnCl)_2O]$, which can be used to convert partially hydrolysed starch formates with C_1-C_5 alkenyl fatty acid esters, e.g. vinyl laurate [247]. Alternative reagents for the transesterification are isopropenyl esters of carboxylic acids, e.g. isopropenyl acetate or N-isopropylidene derivatives such as N-isopropylidene methylcarbonates, ethylcarbonates or benzylcarbonates. Moreover, conversion with methylene diacetate and ethylene diacetate is possible. For cellulose, the transesterification with methylene diacetate and ethylene diacetate is carried out homogeneously in DMSO/paraformaldehyde in the presence of catalytic sodium acetate [190].

5.3.2 Esterification by Ring Opening Reactions

Ring opening reactions of cyclic dicarboxylic acid anhydrides, e.g. succinic, maleic or phthalic anhydrides, are discussed in Chap. 4 because they usually succeed in a similar manner as the synthesis of symmetric anhydrides of aliphatic and aromatic monocarboxylic acids.

An efficient and sophisticated method exploiting a ring opening reaction is the conversion with diketene or with a mixture of diketene/carboxylic acid anhydrides, giving either pure acetoacetates or mixed cellulose acetoacetate carboxylic acid esters of cellulose (Fig. 5.44, [248]).

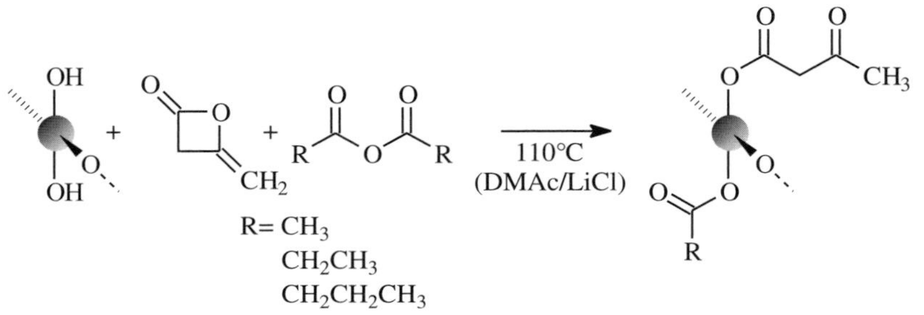

Fig. 5.44. Synthesis of mixed cellulose acetoacetate carboxylic acid esters via conversion with a mixture of diketene/carboxylic acid anhydrides (adapted from [248])

The reaction with diketene is a very useful alternative to the conversion with *tert*-butyl acetoacetate, which does not yield products of high DS values in predictable processes [249, 250]. The reactive intermediate in both cases is acetylketene [251,252]. The reaction can be carried out in DMAc/LiCl or NMP/LiCl. Acetoacetylation with diketene occurs very rapidly at temperatures of $100-110\,°C$, and a complete derivatisation is observed within 30 min (Table 5.38).

Table 5.38. Synthesis of cellulose acetoacetates in DMAc/LiCl with diketene at $110\,°C$ for 30 min, using microcrystalline cellulose (MCC) or hardwood pulp (HWP, adapted from [248])

Conditions			Reaction product
Cellulose Type	Molar ratio		DS
	AGU	Diketene	
MCC	1	0.90	0.78
MCC	1	1.80	1.58
MCC	1	2.70	2.38
HWP	1	0.90	0.84
HWP	1	1.80	1.70
HWP	1	2.70	2.91

Modification of the polymer with mixtures of diketene/carboxylic acid anhydrides gives the same efficiency and predictability as that discussed for the pure ester. Mixed acetoacetate acetates, -propionates and -butyrates can be obtained without catalysis (Table 5.39). This derivatisation gives the polymers solubility ranging from water to THF, depending on the DS of the products. The DP is only negligibly affected during the reactions. The T_g of the cellulose acetoacetates shows no correlation with the DP of the derivative but is strongly influenced by the DS values.

Table 5.39. Synthesis of cellulose acetoacetate (AA) carboxylic acid esters (CE) in DMAc/LiCl with diketene/anhydride (acetic, propionic or butyric) at 110 °C, using microcrystalline cellulose (MCC) or hardwood pulp (HWP, adapted from [248])

Conditions					Reaction Product		
Cellulose Type	Molar ratio				DS		
	AGU	Anhydride		Diketene	CE	AA	Total
HWM	1	Acetic	0.30	0.30	0.20	0.26	0.46
MCC	1	Acetic	0.30	1.50	0.32	1.39	1.71
HWP	1	Acetic	1.50	0.30	1.32	0.37	1.69
HWP	1	Acetic	1.50	1.50	1.07	1.71	2.78
HWP	1	Propionic	0.50	0.50	0.45	0.47	0.92
MCC	1	Propionic	1.50	1.50	1.50	1.40	2.90
HWP	1	Butyric	0.30	0.30	0.22	0.24	0.46
MCC	1	Butyric	1.50	1.50	1.33	1.37	2.70

Ring opening of NMP can be exploited for the preparation of an ionic ester of cellulose. The glucan is converted homogeneously in NMP/LiCl with an intermediate reagent of a Vielsmeier-Haack-type reaction of NMP with TosCl. The reaction procedure yields a cyclic iminium chloride of cellulose. Subsequent hydrolysis of this derivative can follow two possible pathways (Fig. 5.45). One route would regenerate the NMP and cellulose. The other path gives an ester linkage. ^{13}C NMR spectroscopy has revealed that the latter hydrolysis is much faster and forms a cellulose 4-(methylamino)butyrate hydrochloride [253]. ^{1}H,^{1}H-COSY NMR spectroscopy confirms the structural purity of the cellulose ester (Fig. 5.46).

Widely applied is the ring opening of lactones for polysaccharide modification, which usually results in a graft polymerisation. An interesting approach is the synthesis of pullulan derivatives by ring opening of ε-caprolactone and L-lactide, using a tin octanoate catalyst in DMSO (Fig. 5.47). The pullulan 6-hydroxycaproates have DS values between 0.10 and 0.75, as determined by ^{1}H NMR spectroscopy (Table 5.40). The polymers exhibit interesting thermal properties as well as crystallinity [254].

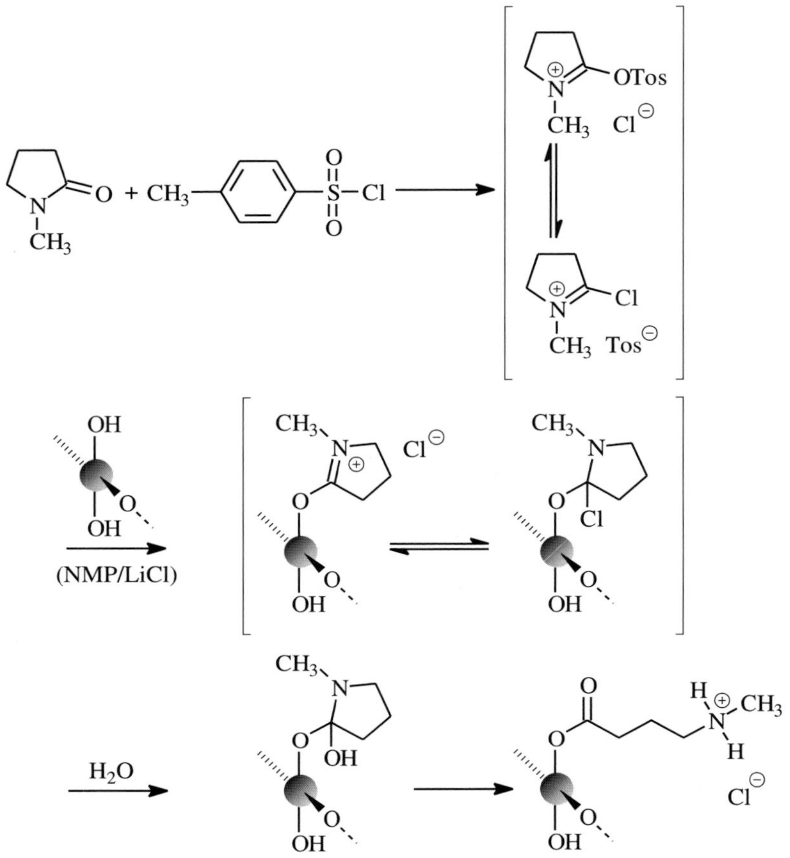

Fig. 5.45. Reaction path for the modification of cellulose with NMP in the presence of TosCl (adapted from [253])

Table 5.40. DS values and solubility of pullulan lactates obtained by ring opening of L-lactide (adapted from [254])

Conditions		Reaction product			
Molar ratio		DS	Solubility		
AGU	Lactide		H_2O	Methanol	Methanol/acetone
1	0.15	0.08	+	–	–
1	0.33	0.21	+	–	–
1	0.66	0.44	–	+	–
1	1.65	0.75	–	–	+

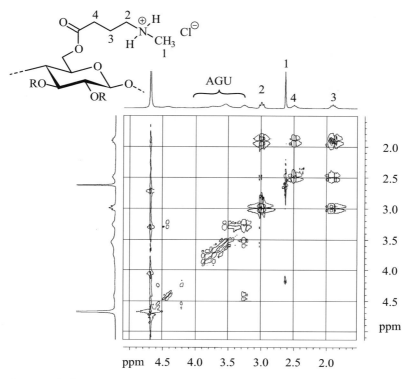

Fig. 5.46. ^{1}H,^{1}H-COSY NMR spectrum of cellulose 4-(methylamino)butyrate hydrochloride obtained by reaction of cellulose with NMP in the presence of TosCl

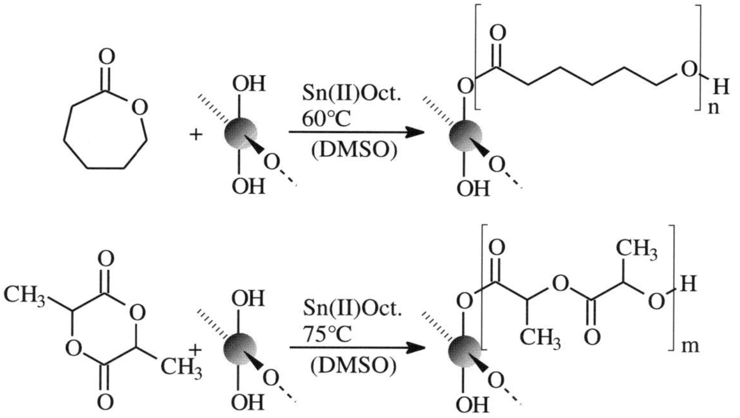

Fig. 5.47. Preparation of pullulan derivatives by ring opening of ε-caprolactone and L-lactide, using a tin-(II)-octanoate catalyst in DMSO (adapted from [254])

In a similar procedure, starch nanoparticles can be acylated with ε-caprolactone in a regioselective manner at position 6, using the immobilised enzyme Novazym 435 and performing the reaction in anhydrous toluene [246]. In the case of cellulose, esterification via ring opening with ε-caprolactone is not successful under heterogeneous conditions. Thus, it is achieved homogeneously in DMAc/LiCl in the presence of TEA with ε-caprolactone at 80 °C for 18 h, giving a DMSO-soluble cellulose 6-hydroxycaproate with a DS of 0.8. More efficient is the reaction in the solvent DMSO/TBAF using tin 2-ethylhexanoate as catalyst (see Fig. 5.48). After 4 h at 60 °C applying 5 mol% catalyst, 100% grafting is achieved. This esterification can also be used for the modification of amylose. A similar reaction in DMSO/tetraethylammonium chloride does not yield modified celluloses. It is assumed that an increase in the nucleophilicity of the hydroxyl groups of cellulose, by the interaction with TBAF, was the cause of this observation [255].

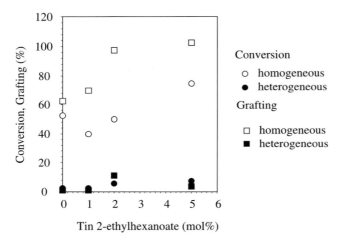

Fig. 5.48. Grafting and conversion achieved depending on the concentration of tin 2-ethylhexanoate (catalyst) for the reaction of cellulose with ε-caprolactone in the solvent DMSO/TBAF (adapted from [255])

It is shown that sulphonic acids and the chlorides are among the useful reagents for the coupling of carboxylic acids onto polysaccharide backbones. The introduction of sulphonic acid moieties is also a valuable synthetic tool, as it delivers the chemistry directly at the carbon atom of the modified RU (as shown in Chap. 6).

6 Sulphonic Acid Esters

Typical structures of sulphonic acid esters used in polysaccharide chemistry are shown in Fig. 6.1. The most widely used are the p-toluenesulphonic- and the methanesulphonic acid esters, due to their availability and hydrolytic stability. The formation of sulphonic acid esters is carried out heterogeneously by conversion with sulphonic acid chlorides in a tertiary organic base, in aqueous alkaline media (NaOH, Schotten-Baumann reaction), or completely homogeneous in a solvent such as DMAc/LiCl. A major drawback of heterogeneous procedures is that long reaction times and a high molar excess of reagent, mostly sulphonic acid chloride, are necessary for significant conversion. Sulphonic acid esters are reactive and may be attacked by unmodified OH groups in situ, yielding cross-links. Hence, the products obtained are insoluble. In addition, they contain a high chlorine content formed by the nucleophilic attack of chloride ions. In contrast, the homogeneous conversion, e.g. of cellulose dissolved in DMAc/LiCl, yields soluble sulphonic acid esters [162]. In particular, homogeneous tosylation applying TosCl in the presence of TEA is very efficient.

It is well known from the chemistry of low-molecular alcohols that hydroxyl functions are converted to a good leaving group by the formation of the corresponding sulphonic acid esters, and hence nucleophilic displacement reactions can be carried out. In the case of polysaccharides, nucleophiles such as halide ions may attack the carbon atom, leading to the corresponding deoxy compound with substitution of the sulphonate group (Fig. 6.2). It is also possible to modify the remaining hydroxyl groups prior to the S_N reaction.

6.1 Mesylates

The heterogeneous conversion of mercerised cellulose (cotton linters) with MesCl (6 mol per mol AGU) in Py slurry affords cellulose mesylate with DS_{Mes} values up to 1.7 after reaction for several days at RT [264]. A crucial point is the activation of the starting cellulose. It was found that the treatment of cellulose with aqueous NaOH increases the reactivity [264]. In order to gain higher DS values, subsequent solvent exchange with anhydrous methanol and Py is necessary. The alkali content of the activated cellulose on the DS_{Mes} is unimportant (Table 6.1).

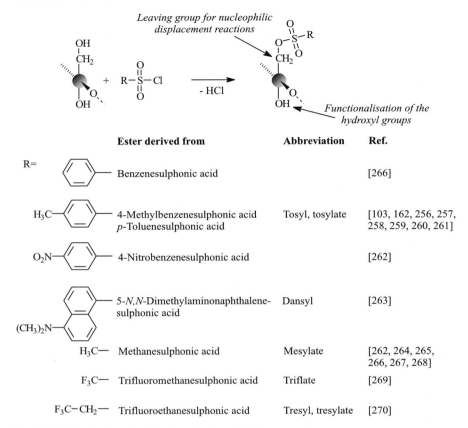

Fig. 6.1. Typical sulphonic acid esters of polysaccharides

Table 6.1. Influence of reaction time and temperature on the conversion of mercerised cellulose (cotton linters) with MesCl in Py (adapted from [265])

Conditions		Reaction product			
Time (h)	Temperature (°C)	S (%)	Cl (%)	DS_{Mes}	DS_{Cl}
4	28	20.0	1.42	2.00	0.11
17	28	23.0	2.91	2.74	0.29
2	57	12.0	10.73	0.93	0.72
20	57	12.5	12.93	1.00	0.86

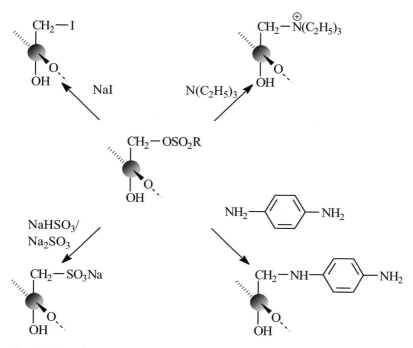

Fig. 6.2. Typical examples for S_N reactions of polysaccharide sulphonic acid esters

The mesylation proceeds readily at 28 °C; at an elevated temperature of 57 °C, side reactions become predominant, yielding products of comparably low DS_{Mes} but with a high content of chlorodeoxy moieties [265].

The conversion of cellulose with MesCl in the presence of TEA in DMAc/LiCl yields products with a DS_{Mes} of 1.3 at a reaction temperature of 7 °C for 24 h (Table 6.2, [266]). By a subsequent mesylation, a DS_{Mes} of 2.1 may be realised [267]. A disadvantage of the homogeneous mesylation is the fact that the reaction has to be carried out at polymer concentrations lower than 1% in order to prevent

Table 6.2. Conversion of cellulose with MesCl homogeneously in DMAc/LiCl at a temperature of 7 °C (adapted from [267])

Conditions Molar ratio				Reaction product			
AGU	MesCl	TEA	Time (h)	S (%)	Cl (%)	DS_{Mes}	DS_{Cl}
1	9	18	24	15.8	1.4	1.3	0.10
1	9	18	48	14.4	2.8	1.2	0.20
1	15	30	24	15.5	1.0	1.3	0.06
1	15	30	48	15.3	0.5	1.3	0.04

gelation during the addition of the reagent. The products have to be precipitated and washed carefully with ethanol/hexane or methanol/acetic acid. The pH value has to be maintained around 7 in order to prevent hydrolysis of the ester moieties. The cellulose mesylates were found to be soluble in DMSO starting at DS_{Mes} 1.3, and additionally in DMF and NMP starting at DS_{Mes} 2.1. Samples with lower DS_{Mes} swell only.

Dextran mesylate is prepared in an aqueous solution of the biopolymer with MesCl and NaOH as base [262]. Precipitation in ethanol yielded a dextran mesylate, which is partly water soluble. The water-soluble fraction (main component) possesses a DS_{Mes} of 0.10, while the DS_{Mes} of the insoluble part is 0.68.

The structure of the reaction products of cross-linked pullulan particles with MesCl depends on the solvent used. In the case of Py, mesylation at 20 °C yields a pullulan mesylate with DS_{Mes} 0.68 and negligible incorporation of chlorodeoxy groups (DS_{Cl} 0.04), applying 3 mol reagent per mol RU. At higher temperatures, S_N reactions become predominant, decreasing the DS_{Mes} and increasing the DS_{Cl}. In contrast, mesylation in DMAc and DMF yields products containing both mesyl and chlorodeoxy moieties already at a low reaction temperature [268]. For instance, if the reaction is carried out in DMF at 20 °C, a DS_{Mes} of 0.04 and a DS_{Cl} of 0.10 were obtained.

6.2 Tosylates

Tosylation of cellulose can be carried out homogeneously in the solvent DMAc/LiCl, permitting the preparation of cellulose qtosylate with defined DS_{Tos} controlled by the molar ratio reagent to AGU at short reaction times, with almost no side reactions [162, 256, 271]. However, the product structure may depend on both the reaction conditions and the workup procedure applied (Fig. 6.3).

Sulphonic acid chloride and DMAc react in a Vilsmeier-Haack-type reaction forming the O-(p-toluenesulphonyl)-N,N-dimethylacetiminium salt I. This intermediate reacts with hydroxyl groups, depending on the conditions applied. Using weak organic bases, e.g. Py (pK_a 5.25) or N,N-dimethylaniline (pK_a 5.15), the reaction with the polysaccharide yields a reactive N,N-dimethylacetiminium salt II. II can form chlorodeoxy compounds III at high temperatures or yields the acetylated polysaccharide IV after aqueous workup. In contrast, stronger bases such as TEA (pK_a 10.65) or DMAP (pK_a 9.70) react with I, yielding a less reactive species V compared with II, and hence lead to the formation of polysaccharide sulphonic acid esters VI without undesired side reactions.

Detailed studies on the preparation of cellulose tosylates demonstrate that various cellulose materials with DP values ranging from 280 to 1020 could be converted to the corresponding tosyl esters [256]. At 8–10 °C, DS_{Tos} values in the range 0.4–2.3, with negligible incorporation of chlorodeoxy groups, were obtained within 5–24 h (Fig. 6.4, Table 6.3).

The cellulose tosylates are soluble in a wide variety of organic solvents. From DS_{Tos} 0.4, they dissolve in aprotic dipolar solvents (DMAc, DMF, and DMSO). The

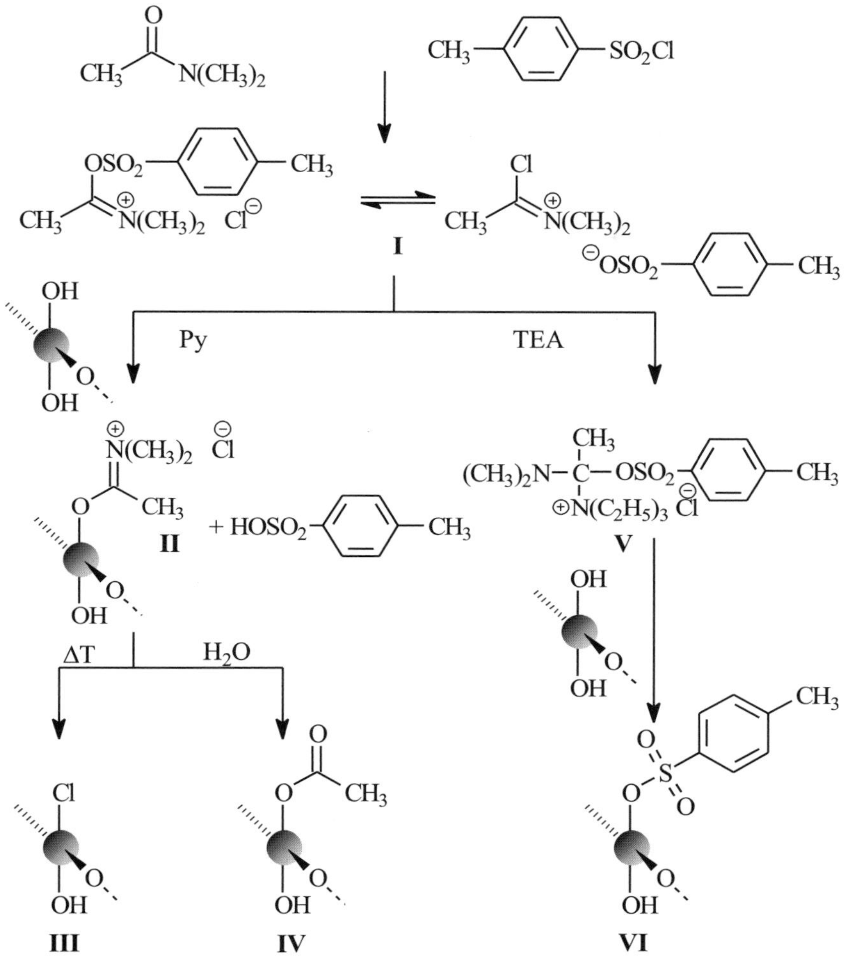

Fig. 6.3. Mechanism for the reaction of cellulose with TosCl in DMAc/LiCl depending on organic bases (adapted from [272])

polymer becomes soluble in acetone and dioxane at DS_{Tos} 1.4 and, in addition, in chloroform and methylenechloride at DS_{Tos} 1.8. A structure characterisation was carried out by means of FTIR- and NMR spectroscopy (Fig. 6.5).

In the ^{13}C NMR spectra (Fig. 6.5), typical signals of the modified AGU are observed in the range from 61.0 to 103.0 ppm. In addition, the peaks of the tosyl ester moiety can be found at 20.5 ppm (methyl group) and in the range from 127.0 to 145.0 ppm (aromatic carbon atoms). It is obvious that position 6 is esterified first because a signal appears at 70 ppm, which is caused by the functionalisation. Significant splitting of the C-1 signal was not detected, indicating that position 2 does not react at low DS_{Tos}. With increasing DS_{Tos}, the intensity of the C-6

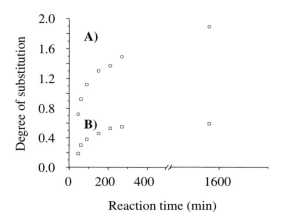

Fig. 6.4. Time dependence of the conversion of cellulose with a molar ratio AGU:TosCl of **A** 1:6 and **B** 1:1 in DMAc/LiCl solution in the presence of TEA

Reaction time (min)

Table 6.3. Reaction of cellulose with TosCl in DMAc/LiCl for 24 h at 8 °C (adapted from [257])

Reaction conditions					Reaction product		
		Molar ratio					
Cellulose	DP	AGU	TosCl	TEA	DS_{Tos}	S (%)	Cl (%)
Microcrystalline	280	1.0	1.8	3.6	1.36	11.69	0.47
		1.0	4.5	9.0	2.30	14.20	0.43
Spruce sulphite pulp	650	1.0	1.8	3.6	1.34	11.68	0.44
		1.0	9.0	18.0	1.84	13.25	0.49
Cotton linters	850	1.0	0.6	1.2	0.38	5.51	0.35
		1.0	1.2	2.4	0.89	9.50	0.50
		1.0	2.1	4.2	1.74	12.90	0.40
		1.0	3.0	6.0	2.04	13.74	0.50
Beech sulphite pulp	1020	1.0	1.8	3.6	1.52	12.25	0.43

peak decreases considerably until almost complete disappearance at DS_{Tos} 1.89. In addition, another signal for C-1 appears (C-1′), which is caused by substituents at position 2.

The heterogeneous conversion of starch with TosCl in Py slurry yields starch tosylates of low DS. In this procedure, the starch is activated by treatment with aqueous Py, followed by solvent exchange. The reactivity of the hydroxyl groups is in the order O-6> O-2> O-3 [258]. Using DMAc/LiCl as solvent in the presence of TEA at 8 °C, pure starch tosylates can be prepared (Table 6.4, [259]). Compared to cellulose, a low LiCl concentration of 1% is sufficient to dissolve the polymer. Starch tosylates with DS_{Tos} values ranging from 0.6 to 2.0 are accessible with chlorine contents lower than 0.42%.

Reactions at room temperature and with increased amount of reagent lead to products with a lower DS_{Tos}, and the chlorine content is remarkably increased

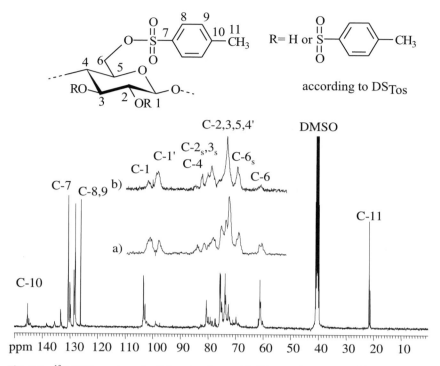

Fig. 6.5. ^{13}C NMR spectra of cellulose tosylates with DS$_{Tos}$ 0.40, **a** 1.12 and **b** 1.89 in DMSO-d_6. The dash (′) means influenced by substitution of the neighbouring position and subscript s means substituted position

Table 6.4. Homogeneous conversion of starch (Hylon VII, 70% amylose) with TosCl in DMAc/LiCl (24 h, adapted from [259])

Reaction conditions				Reaction product		
Molar ratio			Temperature			
AGU	TosCl	TEA	(°C)	DS$_S$	S (%)	Cl (%)
1.0	1.0	2.0	8	0.61	7.61	0.19
1.0	1.5	3.0	8	1.02	10.18	0.20
1.0	2.0	4.0	8	1.35	11.64	0.30
1.0	3.0	6.0	8	1.43	11.93	0.42
1.0	6.0	12.0	8	2.02	13.61	0.32
1.0	1.0	2.0	20	0.61	7.63	0.11
1.0	1.5	3.0	20	0.87	9.40	0.43
1.0	2.0	4.0	20	0.71	8.38	0.45
1.0	3.0	6.0	20	1.27	11.34	1.33
1.0	4.0	8.0	20	1.26	11.30	1.55
1.0	6.0	12.0	20	1.76	12.96	1.48

due to the formation of chlorodeoxy moieties caused by nucleophilic displacement reactions. The starch tosylates are soluble in a variety of solvents. Starting with DS_{Tos} 0.61, they dissolve in aprotic dipolar solvents such as DMAc, DMF and DMSO. The solubility in less polar solvents begins at DS_{Tos} 0.98 in dioxane and at DS_{Tos} 1.15 in THF. A polymer with DS_{Tos} 2.02 can be dissolved in chloroform.

A representative ^{13}C NMR spectrum of a starch tosylate with DS_{Tos} 1.09 is shown in Fig. 6.6. Tosylation leads to a downfield shift of about 8.5 ppm, and hence the carbon atom of the tosylated position C-6 can be found at 69.0 ppm. Additionally, functionalisation of the secondary hydroxyl groups causes signals at 80.2 ppm. It is important to note that an intensive peak appears at 94.3 ppm, which is assigned as C-1', indicating a fully substituted position 2 already at a total DS_{Tos} of 1.09. In contrast to heterogeneous reaction (O-6> O-2/O-3), a reactivity in the order O-2> O-6/O-3 appears.

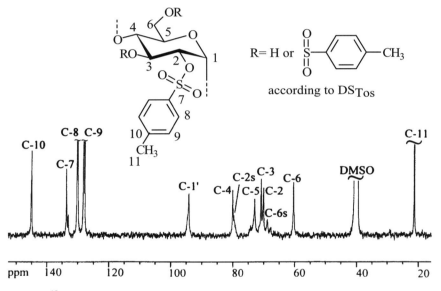

Fig. 6.6. ^{13}C NMR spectrum of starch tosylate with DS_{Tos} 1.09, recorded in DMSO-d_6 at 60 °C (reprinted from Carbohydr Polym 42, Heinze et al., Starch derivatives of high degree of functionalization. 1. Effective, homogeneous synthesis of p-toluenesulfonyl (tosyl) starch with a new functionalization pattern, pp 411–420, copyright (2000) with permission from Elsevier)

Detailed information about the functionalisation pattern of starch tosylates can be obtained by NMR spectroscopy of the peracylated polymers. ^{1}H NMR spectra of peracetylated starch tosylates are shown in Fig. 6.7. The signal at 4.8 ppm is assigned to an acetylated position 2 (H-2), applying two-dimensional NMR methods [273]. The intensity of the H-2 signal decreases with increasing DS_{Tos}. Starting with DS_{Tos} 1.02, the peak disappears. Thus, the tosylation occurs preferably at the hydroxyl group at position 2. This tosylated position undergoes S_N reactions

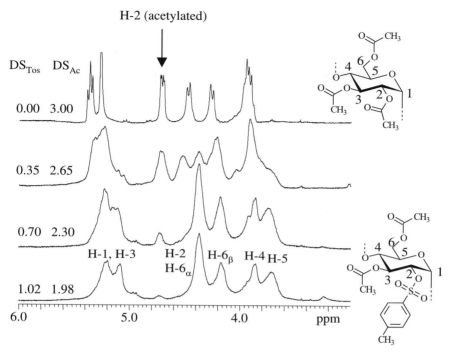

Fig. 6.7. ^{1}H NMR spectra from acetylated starch tosylates of different DS_{Tos} and DS_{Ac}. The spectral range shown (3–6 ppm) is specific for the modified AGU (reprinted from Carbohydr Polym 45, Dicke et al., Starch derivatives of high degree of substitution. Part 2. Determination of the functionalization pattern of *p*-toluenesulfonyl starch by peracylation and NMR spectroscopy, pp 43–51, copyright (2001) with permission from Elsevier)

to a limited extent only. Nucleophilic displacement reactions starting from starch tosylate are not reasonable because a high DS_{Tos} is required to ensure tosylation of the primary OH group.

Chitin possesses two different reaction sites that can be attacked by the sulphonic acid chloride (Fig. 6.8). There are OH groups forming the corresponding ester and NH_2 moieties leading to *N*-sulphonamide, which are not susceptible for S_N reactions. Chitin tosylate, a versatile derivative for subsequent reactions, can be synthesised by homogeneous or heterogeneous processes [103]. Iododeoxy- and mercaptodeoxy chitin are accessible as precursors for, e.g. graft copolymerisation with styrene [274, 275]. In order to prevent *N*-tosylation, applying pyridine as reaction medium and a high excess of TosCl (10 mol/mol repeating unit), products with DS_{Tos} up to 0.83 are obtained as a white fibrous material (Table 6.5).

DMAP promotes the tosylation reaction, and no *N*-deacetylation occurs under these mild conditions. α-Chitin isolated from shrimp is remarkably less reactive, compared with β-chitin from squid pens.

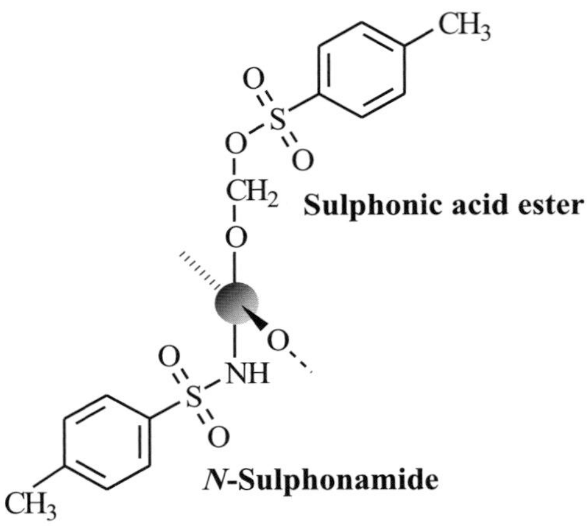

Fig. 6.8. Different types of sulphonic acid derivatives of the 2-deoxy-2-amino repeating unit in chitin

Table 6.5. Preparation of chitin tosylates heterogeneously starting from N-acetyl chitin (0.2 g) in Py in the presence of DMAP (adapted from [103])

Chitin		Reaction conditions		Chitin tosylate
Source	Type	DMAP (g)	Time (h)	DS
Squid	β	0	24	0.34
Squid	β	0	48	0.69
Squid	β	0.2	24	0.62
Squid	β	0.2	48	0.70
Squid	β	1.0	48	0.80
Squid	β	2.0	48	0.83
Shrimp	α	0.2	48	0.18

In a homogeneous procedure, chitin (DDA 0.18) is converted to alkali chitin by treatment with 42% aqueous sodium hydroxide [260]. A solution is obtained after addition of crushed ice. A biphasic mixture is formed with a solution of TosCl in chloroform by vigorously stirring for 2 h at 0 °C and 2 h at 20 °C. After workup, chitin tosylates with DS_{Tos} up to 1.01 were obtained (Table 6.6). The possibility of

Table 6.6. Homogeneous preparation of chitin tosylates starting from alkali chitin (adapted from [260])

Molar ratio		Chitin O-tosylate	
Repeating unit	TosCl	DS_{Tos}	Solubility
1	7	0.42	DMSO, DMAc, NMP, HCOOH,
1	10	0.73	DMSO, DMAc, NMP, HPMA, HCOOH
1	15	0.95	n.d.
1	20	1.01	n.d.

N-tosylation was mentioned to be negligible due to the higher reactivity of the hydroxyl groups compared with NH-functions under strong alkaline conditions. However, acetamide functions may be hydrolysed and acetyl groups may migrate during the tosylation.

It is possible to dissolve chitosan in DMAc containing 5–8% LiCl for tosylation under mild reaction conditions [261]. DS_{Tos} values of up to 1.1 were reached applying 35 mol reagent per mol repeating unit (Fig. 6.9, [276]). The high excess of reagent applied shows that chitin is less reactive than cellulose in the dissolved state.

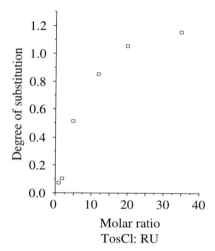

Fig. 6.9. Dependence of the DS_{Tos} of chitin tosylate on the molar ratio TosCl:RU (adapted from [276])

In principal, tosylation of other polysaccharides, e.g. dextran, scleroglucan, xylan and guar, can be carried out by the methods described above. Subsequent modifications of polysaccharide tosylates, such as S_N reactions, are particularly possible if primary hydroxyl functions are present. In the case of polysaccharides that do not contain primary OH functions on the main chain, the S_N reaction is somewhat limited. It has also to be taken into account that nucleophils possess a different reactivity. The azide ion is much more nucleophilic than the iodide ion, and therefore also sulphonic acid ester moieties bound to secondary OH groups can be displaced. In addition, α-(1 → 6) linked polysaccharides (e.g. dextran) may contain also branches with primary OH moieties at the end groups as a deviation from the uniform structure.

Side reactions, such as the formation of C=C double bonds due to elimination reactions, are well known. Although Rogovin et al. [278] reported the tosylation of dextran and subsequent S_N reactions, these conversions have to be revised following analysis by modern spectroscopic techniques from the authors' point of view. The few examples found in the literature are summarised in Table 6.7.

6.3 Miscellaneous Sulphonic Acid Esters

Various other aromatic or aliphatic sulphonic acid chlorides have been reacted with polysaccharides. Substituted benzenesulphonic acid esters of cellulose are investigated in terms of the S_N reaction with lithium acetate, and a reaction mechanism is proposed [277].

Table 6.7. Application of miscellaneous polysaccharide tosylates

Polysaccharide derivative	Application	Reference
Dextran tosylate	Preparation of deoxyiodo dextran and deoxyamino dextran	[278]
Dextran tosylate	S_N reaction with potassium thiocyanate	[279]
Hemicellulose tosylate	S_N reaction with NaSCN	[280]
Scleroglucan tosylate	Introduction of heavy atoms for X-ray fluorescence spectroscopy	[281, 282]

The fluorine-containing aliphatic sulphonic acid esters of polysaccharides exhibit a high reactivity and have to be handled carefully. The conversion of trifluoromethanesulphonic acid anhydride with methyl cellulose and cellulose acetate was used for the preparation of cross-linked gels [269]. 2,2,2-trifluoroethanesulphonic acid chloride was found to be a useful reagent for improving the affinity for cellulose acetate membranes for enzyme immobilisation [270]. Starting from cellulose 2,2,2-trifluoroethanesulphonic acid ester, deoxyaminocelluloses with chromophoric properties as carrier matrices were prepared [283]. Cellulose esters bearing aminoaryl sulphonic acid may have useful optical properties (fluorescence) [263].

7 Inorganic Polysaccharide Esters

Polysaccharides form esters with any inorganic acid known. Examples of typical products are summarised in Fig. 7.1. The esters of nitric acid, phosphoric acid, dithiocarbonic acid and sulphuric acid have gained importance. Cellulose nitrate is commercially produced and used as, for example, film-forming component in lacquers and as explosive. However, the inorganic esters of cellulose and other polysaccharides have yet to be commercially exploited. Anionic functions such as sulphuric acid half esters are found in numerous naturally occurring polysaccharides. Typical examples are heparan and chondroitin [284].

Esters of polysaccharides with functional groups that can be split off by changing the conditions (pH value, medium, salt concentration) are used for shaping processes. The most important commercial example is the 3 000 000 t annual worldwide production of rayon via cellulose dithiocarbonic acid ester (xanthogenate). The cellulose xanthogenate is formed by treating cellulose with CS_2/NaOH, and dissolves in the surplus of aqueous NaOH during xanthogenation. The viscose process is described in detail in [285]. The conversion of polysaccharides with N_2O_4 in the presence of a polar aprotic solvent under dissolution yields the nitrite, which can be used for regeneration by applying a protic solvent [286, 287].

7.1 Sulphuric Acid Half Esters

Polysaccharides containing sulphuric acid half ester moieties constitute a complex class of compounds occurring in living organisms. They possess a variety of biological functions, e.g. inhibition of blood coagulation, or are a component of connective tissues [288]. These polysaccharides are usually composed of different sugars including aminodeoxy- and carboxylic groups containing RU, e.g. β-D-glucuronic acid or α-L-iduronic acid and N-acetyl-β-D-galactosamine [289].

Heparan sulphate is composed of α-L-iduronic acid and N-acetyl-β-D-galactosamine (Fig. 7.2A, [290]). The structure of heparin is similar to that of heparan sulphate but it contains higher amounts of sulphate groups and iduronic acid. The sulphate ester moieties are bound to position 3 of the L-iduronic acid and position 6 of the D-galactosamine. Moreover, the amino group is either acetylated or sulphated. Heparin is an important therapeutic anticoagulant and antithrombotic agent.

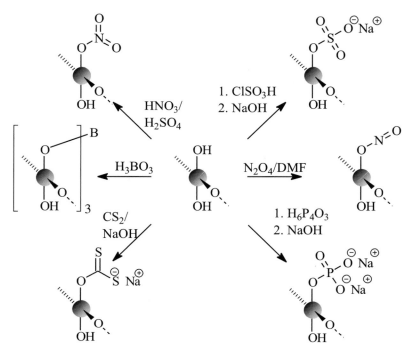

Fig. 7.1. Examples of polysaccharide esters of inorganic acids

The main sugar moieties of chondroitin, a component of cartilage and connective tissue, are β-D-glucuronic acid and N-acetyl-β-D-galactosamine connected via β-(1→3) linkages [291]. Sulphuric acid half esters are found at position 4 or 6 of the N-acetyl-β-D-galactosamine (see Fig. 7.2B for chondroitin-6-sulphate). Dermatan sulphate consists of L-iduronic acid, rather than D-glucuronic acid (Fig. 7.2C) [292].

Sulphuric acid half ester moieties are introduced in polysaccharides in order to render water-insoluble biopolymers soluble and to impart biological activity. For instance, curdlan, which is not very water soluble, gives clear solutions after introduction of a small amount of sulphuric acid half ester groups, as little as 4.4 mol% (DS 0.04) [293]. Consequently, sulphation of polysaccharides is an important path for structure- and property design.

Several homogeneous and heterogeneous synthesis paths have been developed for the preparation of artificially sulphated polysaccharides. The ester, in its H^+ form, is strongly acidic, which causes autocatalytic hydrolysis of the ester moieties and also chain degradation. Therefore, it is converted to the salt form, often the sodium salt, which is water soluble and stable in aqueous systems.

In general, sulphation can be accomplished using various reagents such as $ClSO_3H$, SO_3 and H_2SO_4. Treating polysaccharides with concentrated or slightly diluted H_2SO_4 may lead to sulphation. Under these conditions, a remarkable

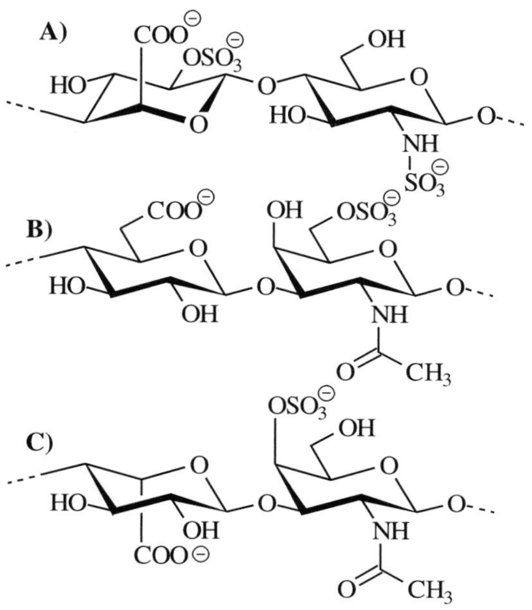

A)

B)

C)

Fig. 7.2. Typical repeating units of heparan sulphate (**A**), chondroitin-6-sulphate (**B**), and dermatan sulphate (**C**)

depolymerisation occurs. H_2SO_4 can also be applied in combination with low-molecular alcohols because alkyl sulphates are formed and act as reactive species. In addition, the polymer degradation is comparably low. Chlorosulphonic acid and sulphur trioxide are powerful sulphating agents, although a major drawback of these reagents is the sensitivity to moisture. A convenient method to reduce the risk during the synthesis is the application of the complexes of $ClSO_3H$ and SO_3 with organic bases (e.g. TEA, Py) or aprotic dipolar solvents (e.g. DMF), which are commercially available. SO_3-DMF and SO_3-Py are white solids that are easy to use. These efficient and easily manageable reagents produce well-defined polysaccharide sulphuric acid half esters, which may exhibit bioactivity. Curdlan-and sulphuric acid half esters are in the centre of interest as cancerostatics and anti-HIV agents.

DMF is a typical reaction medium for the sulphation of polysaccharides, e.g. amylose and amylopectin, which dissolves or at least swells the polymer. It can also be applied with comparable efficiency for guaran, as shown in Fig. 7.3 [294].

A convenient method for the synthesis of curdlan sulphuric acid half ester is dissolution of the polymer in aprotic dipolar media, treatment with SO_3-Py, and subsequent neutralisation to the sodium salt. The sulphation of curdlan swollen in formamide with SO_3-Py complex yields products with DS as high as 2.10 within 4 h at RT (Table 7.1, [295]).

Curdlan sulphuric acid half esters with DS 1.6 are obtainable using piperidine-N-sulphonic acid in DMSO solution. Sulphation with SO_3-Py complex in Py slurry yields products with DS up to 2.6, while almost complete functionalisation can be

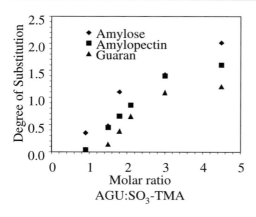

Fig. 7.3. Sulphation of polysaccharides in DMF with SO_3-TMA for 24 h at 0 °C (adapted from [294])

Table 7.1. Sulphation of curdlan with SO_3-Py in formamide for 24 h at RT (adapted from [295])

Molar ratio		Product				
AGU	SO_3-Py			Partial DS		
		S (%)	DS	O-6	O-4	O-2
1	2.0	10.4	0.93	n.d.	n.d.	n.d.
1	2.5	13.6	1.45	1.00	0.14	0.31
1	3.0	14.6	1.75	1.00	0.31	0.44
1	4.0	15.6	1.98	1.00	0.50	0.48
1	5.0	16.1	2.10	n.d.	n.d.	n.d.

achieved with $ClSO_3H$ in Py. The latter two products possess negative specific rotations, leading to the conclusion that the original helical structure of curdlan might be retained. According to ^{13}C NMR measurements, the piperidine-N-sulphonic acid is highly O-6 selective, while OH groups at position 6 still remain in the products prepared with SO_3-Py and $ClSO_3H$ (Table 7.2, [296]).

Table 7.2. Sulphation of curdlan with different reagents for 60 min at 85 °C

Reaction conditions				Curdlan sulphate	
Sulphating reagent	Molar ratio		Temp. (°C)	DS	$[\alpha]_D^{25}$ (°)
	AGU	Reagent			
Piperidine-N-sulphonic acid	1	4.0	85	1.60	−1.5
SO_3-pyridine complex	1	15.2	85	2.40	−14.9
$ClSO_3H$	1	6.3	100	2.60	−16.0
$ClSO_3H$	1	12.6	100	2.90	−22.0

Conversion of curdlan dissolved in DMSO/LiCl with SO_3-Py for 4 h at 80 °C yields sulphuric acid half esters with DS 1.7 [177]. Highly derivatised polysaccharide sulphuric acid half esters can be prepared using a large excess of the reagent in DMF for 6 h at 40 °C. For instance, a DS of 2.8–2.9 can be realised for curdlan, galactan, amylose and cellulose. DS 2.1 is achieved for xylan, which contains not only xylose RU but also some glucuronic acid moieties [297].

A heterogeneous procedure for the sulphation of $(1{\rightarrow}3)$-β-D-glucans isolated from *Saccharomyces cerevisae* with a H_2SO_4/isopropanol mixture yields the sulphuric acid half ester [298]. After the reaction, the product is filtered off and separated from unreacted starting polymer by dissolution in water. The yield (37%) is low, compared to homogeneous conversions.

Sodium alginate sulphuric acid half ester is obtainable by the reaction of sodium alginate with a mixture of $ClSO_3H$ and formamide. The product obtained after 4 h at 60 °C possesses a DS of 1.41 [299]. Sodium alginate sulphuric acid half ester shows a considerably high anticoagulant activity, which can be further increased by subsequent quaternisation with 2,3-epoxypropyltrimethylammonium chloride.

Starch sulphuric acid half esters have been prepared by using various reagents (Table 7.3). Highly sulphated products with DS as high as 2.9 are obtainable with $ClSO_3H$, SO_3, SO_3-Py, SO_3-DMSO in aprotic media. Interestingly, sulphation can be achieved also in aqueous reaction media applying different reagents derived from SO_3 or $ClSO_3H$. Sulphation to low DS values is carried out in order to improve the dissolution behaviour of starch.

Table 7.3. Reagents for the sulphation of starch

Reagent	Medium	DS	Ref.
H_2SO_4	Diethylether	0.25	[300]
$ClSO_3H$	Formamide		
SO_3/formamide		< 1.8	[301–304]
Sodium nitrite/sodium bisulphite	Dry	< 0.02	[305]
Urea/sulphamic acid	Dry	< 0.04	[306]
SO_3/tertiary amine	Water/organic solvent		[307, 308]
SO_3/Py	DMF	2.9	[297]
SO_3/DMSO	DMSO	2.4	[309]
SO_3/DMF	DMF	1.6	[310, 311]
SO_3/TMA	H_2O	< 0.1	[312]
N-Methylimidazole-N'-sulphonate	H_2O	< 0.01	[313]

Many procedures have been developed for the preparation of cellulose sulphuric acid half esters. Simple dissolution of cellulose in 70–75% aqueous H_2SO_4 yields a sulphated but highly degraded derivative. Heterogeneous synthesis, using a mixture of H_2SO_4 and low-molecular alcohols, e.g. *n*-propanol, leads to cellulose sulphuric acid half ester with DS $\approx$ 0.4. The intermediately formed propylsulphuric acid half ester acts as sulphating reagent. The H_2SO_4/n-propanol mixture can be easily reused [314].

Many of the sulphating reagents are highly reactive and, hence, the substituents are not uniformly distributed along the polymer chain. This may render the products water insoluble, even at high DS. The sulphation of dissolved cellulose can yield a uniform functionalisation pattern. However, DMAc/LiCl, for example, is not the solvent of choice for sulphation, because insoluble products of low DS are obtained [315].

Although N_2O_4/DMF is a hazardous cellulose solvent, it is very useful for the preparation of cellulose sulphuric acid half esters. The intermediately formed nitrite is attacked by various reagents (SO_3, $ClSO_3H$, SO_2Cl_2, H_2NSO_3H), leading to cellulose sulphuric acid half esters via transesterification, with DS values ranging from 0.3 to 1.6 after cleavage of the residual nitrite moieties [192,316]. The regioselectivity of the transesterification reaction can be controlled by reaction conditions (Table 7.4). In contrast to the direct sulphation of cellulose, the polymer degradation is rather low, leading to products that form highly viscous solutions. The residual nitrite moieties are cleaved during the workup procedure under protic conditions.

Table 7.4. Sulphation of cellulose nitrite with different reagents (2 mol/mol AGU). The DS values were determined by means of NMR spectroscopy (adapted from [192])

Reaction conditions			Reaction production			
Reagent	Time (h)	Temp. (°C)		Partial DS		
			DS	O-2	O-3	O-6
$NOSO_4H$	4	20	0.35	0.04	0	0.31
NH_2SO_3H	3	20	0.40	0.10	0	0.30
SO_2Cl_2	2	20	1.00	0.30	0	0.70
SO_3	3	20	0.92	0.26	0	0.66
SO_3	1.5	−20	0.55	0.45	0	0.10

In order to circumvent the use of toxic N_2O_4/DMF, cellulose derivatives with activating substituents are useful starting materials. A typical example is TMS cellulose, which is soluble in various solvents, e.g. DMF and THF, and readily reacts with SO_3-Py or SO_3-DMF [317]. The preparation of TMS cellulose is quite simple and can be achieved by heterogeneous conversion of cellulose in DMF/NH$_3$ with trimethylchlorosilane (DS $< \approx$ 1.5) or homogeneously in DMAc/LiCl with hexamethyldisilazane. The latter method has been used for the preparation of TMS celluloses with DS up to 2.9.

As in the sulphation of cellulose nitrite, the TMS group acts as leaving group. The first step consists of an insertion of SO_3 into the Si–O bond of the silyl

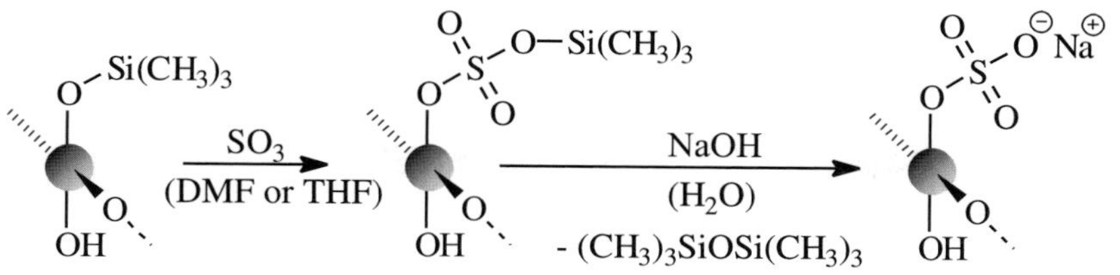

Fig. 7.4. Preparation of cellulose sulphate via trimethylsilyl cellulose

ether (Fig. 7.4). The intermediate formed is unstable and usually not isolated. Subsequent treatment with aqueous NaOH leads to a cleavage of the TMS group under formation of the sodium cellulose sulphuric acid half ester.

Due to the course of reaction the DS_S is limited by the DS_{Si} of the starting TMS cellulose and can be adjusted in the range from 0.2 to 2.5. Typical examples are summarised in Table 7.5. The sulphation reaction is fast and takes about 3 h, with negligible depolymerisation. Thus, products of high molecular mass are accessible if a TMS cellulose of high DP is applied as starting material. For instance, the specific viscosity of a cellulose sulphuric acid half ester with DS 0.60 is 4900 (1% in H_2O, [317]).

Table 7.5. Sulphation of cellulose via TMS cellulose (adapted from [317])

Reaction conditions					Product
TMS cellulose	Solvent	Sulphating agent			DS
DS_{Si}		Type	Molar ratio		
			AGU	Reagent	
1.55	DMF	SO_3	1	1.0	0.70
1.55	DMF	SO_3	1	2.0	1.30
1.55	DMF	SO_3	1	6.0	1.55
1.55	DMF	$ClSO_3H$	1	1.0	0.60
1.55	DMF	$ClSO_3H$	1	2.0	1.00
1.55	DMF	$ClSO_3H$	1	3.0	1.55
2.40	THF	SO_3	1	1.0	0.71
2.40	THF	SO_3	1	1.7	0.90
2.40	THF	SO_3	1	3.3	1.84
2.40	THF	SO_3	1	9.0	2.40

The sulphation can be carried out in a one-pot reaction, i.e. without isolation and redissolution of the TMS cellulose [317]. Thus, after the silylation of cellulose in DMF/NH$_3$, the excess NH$_3$ is removed under vacuum, followed by separation of the NH$_4$Cl formed. The sulphating agent, e.g. SO$_3$ or ClSO$_3$H, dissolved in DMF is added and the cellulose sulphuric acid half ester is isolated.

Cellulose sulphuric acid half esters of low DS are used for the preparation of symplex capsules. In the case of cellulose, sulphuric acid half ester with a DS as low as 0.2 is sufficient to impart water solubility if the substituents are uniformly distributed along the polymer chain. This can be easily realised by sulphation of a commercially available cellulose acetate with DS 2.5 in DMF solution [186]. The acetyl moieties act as protecting group, and the sulphation with SO_3-Py, SO_3-DMF or acetylsulphuric acid proceeds exclusively at the unmodified hydroxyl groups (Fig. 7.5). No transesterification occurs. The cellulose sulphuric acid half ester acetate formed is neutralised with sodium acetate and subsequently treated with NaOH in ethanol as slurry medium to cleave the acetate moieties. In order to decrease polymer degradation, the saponification is carried out in an inert atmosphere for 16 h at room temperature.

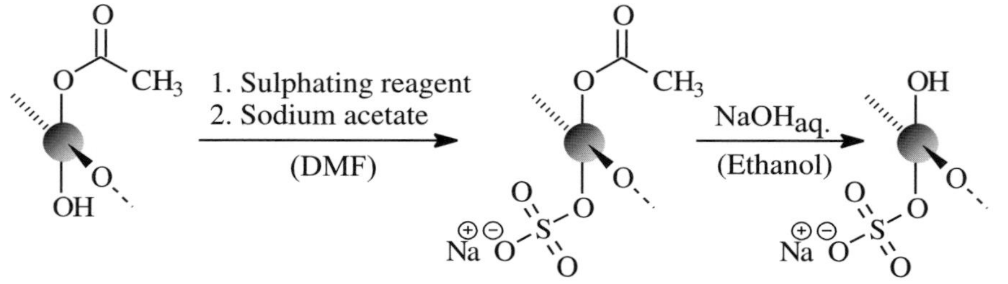

Fig. 7.5. Preparation of cellulose sulphuric acid half ester starting from cellulose acetate, acetyl moieties acting as protective groups

7.2 Phosphates

The introduction of phosphate groups into sugar molecules is an important activation step in the biosynthesis of polysaccharides. The phosphate moieties are split off during the polysaccharide formation and only a small phosphorous content remains in the polymer. In the case of starch, about 0.1% P as phosphoric acid monoester may exist [318]. The amount of phosphate moieties bound to the starch backbone depends on the starch source and has a major impact on the rheological properties [319]. Starch phosphorylation plays an important role in metabolism, as reviewed in [320].

Phosphoric acid is trifunctional and possesses the ability to cross-link the polysaccharide, which can lead to insoluble products with undefined structure. Phosphorylation is carried out in order to retard the dissolution of polysaccharides or to impart biological activity. Phosphorylation increases the flame retardancy of textile fibres. Cellulose phosphates may be used as weak cation exchangers. For this purpose, insolubility in aqueous media is required.

Starch phosphates, which are widely used as food additives, have been extensively studied in order to control the rheological behaviour. Starch phosphates have also found application as wet-end additives in paper making and for adhesives in textile production, where products of low DS (usually $\approx 0.4\%$ PO$_4$ groups) are used. Products with up to 12% phosphate groups are applied in agriculture and pharmaceuticals.

The introduction of phosphoric acid ester moieties can be accomplished by means of different reagents such as polyphosphoric acid, POCl$_3$, P$_2$O$_5$ and phosphoric acid salts. The acid form of the ester is mostly transformed to the alkali salt. Moreover, phosphorylation reagents, summarised in Table 7.6, especially the phosphororganic compounds, enable the preparation of monoesters of low DS under heterogeneous conditions in aqueous media.

Table 7.6. Reagents for the preparation of starch phosphate monoesters

Reagent		Ref.
[structure: sodium tripolyphosphate]	Sodium tripolyphosphate	[321–323]
[structure]	N-Phosphoryl-N'-methylimidazole	[324]
[structure]	N-Benzoylphospho-amidic acid	[325]
[structure]	Salicylphosphate	[326]
CH$_2$=CH–P(=O)(OC$_2$H$_5$)OC$_2$H$_5$	Diethylvinylphosphonate	[327]
[structure]	N-Phosphoryl-2-alkyl-2-oxazoline	[328]

Sodium tripolyphosphate and $Na_3HP_2O_7$ (sodium pyrophosphate) react with starch under formation of the phosphates (DS 0.02) [321–323]. Although temperatures of 100–120 °C are usually required, the reaction with orthophosphates, e.g. NaH_2PO_4/Na_2HPO_4 has to be carried out at higher temperatures (140–160 °C) but yields increased DS values reaching 0.2 [323, 329].

In addition to the simple mixing of starch with inorganic phosphates and subsequent "baking", the biopolymer can be impregnated with aqueous solutions of the phosphorylation reagents. Some is retained in the polymer, which is separated from the solution and heat-treated at 140–160 °C [330]. A starch phosphate with DS 0.14 is obtained by treating potato starch with Na_2HPO_4/NaH_2PO_4 (2.5 mol/mol AGU) in H_2O for 20 min at 35 °C (pH 6), followed by filtration, drying and heat treatment for 3 h at 150 °C [331]. Generally, the preparation of starch phosphates by means of a slurry process is more efficient than dry mixing and heating [332]. Introduction of phosphate groups decreases the gelatinisation temperature and improves the freeze-thaw stability of modified starch-containing mixtures. The "baking" process of the impregnated starch is carried out in, for example, rotating drums, fluidised bed reactors or extruders. Energy input by means of ultrahigh-frequency irradiation at 2450 MHz may be used (for details, see [333, 334]).

The polysaccharide monophosphate tends to cross-linking. It is therefore important to control the pH value of the phosphorylation mixture; in particular, pH 5.0–6.5 is optimal for reactions with orthophosphates while sodium tripolyphosphate can be converted between pH 5.0 and 8.5 [335]. At higher pH, cross-linking via formation of diesters may become predominant [336].

The simultaneous reaction of starch with inorganic phosphates and urea was found to be effective due to synergistic effects, which provides access to modified starch with higher viscosity and less colour [337]. It has to be taken into account that the products contain a certain amount of nitrogen. A typical preparation consists in the processing of a starch containing 5% NaH_2PO_4 and 10–15% urea. The resulting material contains 0.2–0.4% N and 0.1–0.2% P [338]. An additional heat treatment for several hours at 150 °C under vacuum (50–500 Torr) leads to starch phosphates with 0.31–2.1% P and 0.08–0.5% N [339]. In the case of potato starch, the P content of the native material can be increased from 2.04 to 3.07 (DS 0.056) with no change in the granular structure [340]. The reaction conditions strongly influence the properties of the products [341].

Starch can also be phosphorylated applying organophosphates. For example, the reaction of 2,3-di-O-acetylamylose with dibenzylchlorophosphate yields a product with DS 0.7 after cleavage of the acetyl and benzyl moieties [342, 343]. Higher DS values (1.75) can be realised applying tetrapolyphosphoric acid in combination with trialkylamine in DMF for 6 h at 120 °C [343, 344]. A highly reactive phosphorylation reagent is $POCl_3$ in DMF, yielding water-soluble products with up to 11.3% P. Cross-linking may occur due to the three reactive sites.

Curdlan phosphate can be prepared by treatment of the polysaccharide with phosphoric acid, their salts, or $POCl_3$. It is accomplished by the impregnation of curdlan with aqueous solutions of phosphate salts, drying and heat treatment at

120 °C yielding a product with 0.99% P [345], i.e. the procedure is similar to the phosphorylation of starch.

Guar phosphate is prepared by impregnation of the polymer with aqueous NaH_2PO_4 for 30 min at 60–66 °C (pH 6), drying and heat treatment at 150–160 °C (DS values not given). The polymer dissolves in water and the viscosity of a 2% solution is 400–500 cP. Moreover, cross-linking with Al ions is possible, indicating the polyelectrolyte behaviour [346]. It was also found that polysaccharide phosphates exhibit biological activity. Dextran is treated with polyphosphoric acid in formamide for 48 h, yielding products with up to 1.7% P exhibiting immunostimulatory effects. The mitogenic response of murine splenocytes was enhanced. This effect is independent of the molecular mass [347].

The reaction of chitin with P_2O_5 in methanesulphonic acid yields chitin phosphates [348]. Almost complete substitution is realised applying 4 mol P_2O_5 per mol repeating unit. Chitin phosphates are water soluble independently of the DS. In contrast, phosphates of high DS prepared starting from chitosan (obtained by deacetylation of chitin) are water insoluble due to their amphoteric nature. Cross-linking can be achieved with adipoyl chloride in methanesulphonic acid. The insoluble chitin phosphates are capable of binding metal ions [349]. The conversion of the biopolymer with P_2O_5 in methanesulphonic acid can be applied for the phosphorylation of xylan (22–45% phosphate). It is worth noting that the anticoagulant activity increases with increasing molecular mass and is inversely related to the phosphate content [350].

Various reagents have been examined for cellulose phosphorylation. Anhydrous H_3PO_4 can be used on its own, or in combination with P_2O_5 [351]. The conversion of cellulose with molten H_3PO_4 and urea at 120 °C yields products (DS 0.3–0.6) possessing high water retention values with 0.1–0.2% N [352]. Phosphorylation of cellulose can be achieved by heat treatment of a mixture of cellulose/urea/H_3PO_4 at 150 °C [353]. A mixture of H_3PO_4, P_2O_5 and $(C_2H_5O)_3PO$ in hexanol yields a cellulose phosphate gel (maximum DS 2.5) that is expected to promote calcium phosphate formation during bone regeneration [354].

The effective reagent $POCl_3$ can be used in different reaction media (e.g. DMF, Py) but yields only partly soluble products due to cross-linking. Moreover, if the reaction is carried out in DMF, chlorination occurs due to the formation of iminium compounds [355], and accordingly formamide is the preferred reaction medium for conversions of cellulose with PCl_3, PCl_5 and $POCl_3$ [356]. The reaction of cellulose dissolved in non-derivatising solvents (e.g. DMAc/LiCl) does not yield soluble products. Spontaneous gelation occurs, resulting in inhomogeneous products [357]. N_2O_4/DMF is a good solvent for the preparation of cellulose phosphates. Products with DS of up to 0.8 can be obtained by using 6 mol $POCl_3$ and 6 mol TEA per mol AGU for 4 h at 50 °C. A large excess of reagent leads to a fast gelation of the reaction mixture, combined with decrease of the reagent yield. Chlorination can be minimised by adding TEA followed by $POCl_3$/DMF. Products with DS 0.3–0.6 are water soluble [358]. $POCl_3$ can be partially hydrolysed with water or treated with N_2O_4 in order to reduce the reactivity of the reagent [359]. The water solubility of the cellulose phosphates strongly depends on the workup conditions, where cross-links of diester moieties may be hydrolysed. Water-soluble cellulose phosphates can also be prepared starting from cellulose acetates with polytetraphosphoric acid and subsequent deacetylation [360].

7.3 Nitrates

Polysaccharide nitrates are most commonly prepared by treatment of the polysaccharides with mixtures of nitric acid and sulphuric acid. The reaction of inulin with a mixture containing 48.5% HNO_3 and 51.2% H_2SO_4 yields inulin nitrate (13.75% N), while the product obtained with a mixture of 20.8% HNO_3 and 62.8% H_2SO_4 contains 12.8% N [361]. Treatment of dextran with HNO_3 containing $\geq 38\%$ H_2SO_4 as well as HNO_3 with 25% acetic acid and 25% acetic anhydride leads to dextran nitrate [362]. The conversion of chitosan (DDA 85%) with absolute HNO_3 or mixtures of absolute HNO_3 with acetic acid and acetic anhydride gives chitosan nitrates with DS up to 1.7. The amino group of the chitosan nitrate is protonated during the reaction, and isolated with NO_3^- as counter ion. It is possible to substitute NO_3^- by ClO_4^- ions. However, this derivative is unstable and tends to spontaneous decomposition during storage [363].

Starch nitrates with almost complete functionalisation (DS 3) can be used as explosives [364]. The manufacturing procedure is similar to the nitration of cellulose, i.e. the starch is treated with a nitrating mixture containing 32.5% HNO_3, 64.5% H_2SO_4 and 3% H_2O in a solid:liquid ratio of 25:100 [333]. Dissolving starch in N_2O_4/DMF and subsequent conversion of the intermediately formed nitrite by heating in the presence of methanol leads to starch nitrate [286, 316].

Cellulose nitrate is by far the most important polysaccharide ester of nitric acid. Cellulose nitrate, often misnamed "nitro cellulose", is used in many application fields (Table 7.7). Nitration is achieved by using different reagents allowing the control of DS (Fig. 7.6).

Table 7.7. Solubility and application of cellulose nitrate in function of the DS

Nitrogen content (%)	DS	Solubility	Application
11.8–12.2	2.20–2.32	Esters, ketones, ether-alcohol mixtures	Industrial coatings
10.9–11.2	1.94–2.02	Ethanol, isopropanol	Plastic foils, flexographic inks
12.6–13.8	2.45–2.87	Esters	Explosives

It can be easily prepared by treating cellulose, e.g. cotton linters, with nitrating acid containing a mixture of H_2SO_4 and HNO_3. The DS can be controlled by adjusting the composition of the nitrating acid with varying amounts of the dewatering agent H_2SO_4 (Table 7.8).

Fig. 7.6. Reagents for the preparation of cellulose nitrate of different DS

R= H or NO$_2$ depending on the DS

Nitrating agent	DS	Ref.
HNO$_3$/H$_2$SO$_4$	up to 2.9	[365]
HNO$_3$ (90%)/H$_3$PO$_4$/P$_4$O$_{10}$	3.0	[366]
HNO$_3$ (90%)/(CH$_3$CO)$_2$O	3.0	[367]
HNO$_3$ (>98%)/CH$_2$Cl$_2$	2.9	[368]
N$_2$O$_4$/DMF	0.2-0.6	[369, 370]

Table 7.8. Dependence of the DS of cellulose nitrate on the composition of the nitrating acid (adapted from [371])

| Composition of the nitrating acid (%) | | | Cellulose nitrate | |
H$_2$O	HNO$_3$	H$_2$SO$_4$	Nitrogen content (%)	DS
12.0	22.0	66.0	13.2	2.6
16.0	20.0	64.0	12.5	2.4
20.0	20.0	60.0	10.6	1.9

Using HNO$_3$/H$_2$SO$_4$ mixtures, the maximum DS is limited to 2.9 because of the introduction of cellulose sulphuric acid half esters in a competing reaction. These ester moieties are the reason for the instability of crude cellulose nitrate. Careful washing with boiling water at controlled pH yields stable cellulose nitrate.

Complete nitration (DS = 3) can be achieved using nitric acid (90%) in combination with phosphoric acid and phosphorous pentoxide as water-binding agent [366] or a mixture of nitric acid (90%) and acetic acid anhydride [367]. The acetyl nitrate formed acts as highly reactive nitrating agent. Water-free HNO$_3$ in combination with methylene chloride or chloroform yields cellulose nitrates with DS 2.87–2.94 of high stability [368]. Cellulose nitrate with DS 3 is used for analytical purposes, i.e. the determination of the molecular mass and molecular mass distribution by means of viscometry and SEC. Under water-free conditions, the nitration does not affect the DP of the polymer.

8 Structure Analysis of Polysaccharide Esters

For modified polysaccharides, the analysis goes far beyond the structural verification. The chemical structure of the ester function introduced, the DS, and the distribution of the functional groups at both the level of the RU and along the polymer backbone (Fig. 8.1) can strongly influence the properties and need to be determined comprehensively.

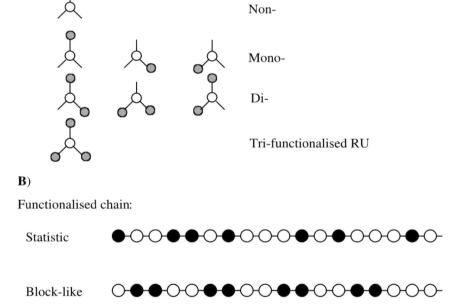

Fig. 8.1. Schematic plot of the possible patterns of functionalisation for the repeating units (**A**) and for the distribution along the polymer chain (**B**) of polysaccharides with three reactive sites

The chemical functionalisation may be associated with side reactions modifying the polymer backbone additionally, maybe to a rather low extent only. These "structural impurities" introduced have to be revealed as well because they are not removable from the polymer chain. Consequently, an efficient and reliable analysis (type of functionalisation, DS, the pattern of substitution) is indispensable for the establishment of structure–property relations of the modified polymers. The tailored modification of substitution patterns can be used to "fine tune" product properties, e.g. solubility behaviour, as shown for the water solubility of cellulose acetate in Table 8.1 [89].

Table 8.1. Water solubility of cellulose acetate: dependence on the pattern of functionalisation (adapted from [89])

Method[a]	Total DS[b]	Degree of acetylation at position[b]			Water-soluble fraction (%, w/w)
		2	3	6	
1	0.49	0.16	0.13	0.20	29
1	0.66	0.23	0.20	0.23	99
1	0.90	0.31	0.29	0.30	93
2	0.73	0.18	0.19	0.36	30
3	1.10	0.33	0.25	0.52	5

[a] Methods applied: (1) deacetylation of cellulose triacetate with aqueous acetic acid, (2) reaction of cellulose triacetate with hydrazine, and (3) acetylation of cellulose with acetic anhydride in DMAc/LiCl

[b] Determined by ^{13}C NMR spectroscopy

Analytical data are also necessary to confirm the reproducibility of a synthesis and the resulting product purity. Unconventional polysaccharide esters, e.g. with sensitive heterocyclic moieties, can often not be analysed by "standard methods" and this has required the development of new analytical tools.

Most of the structural features of the polymer backbone are accessible via optical spectroscopy, chromatography and NMR spectroscopy, as discussed in Chap. 3. Specific techniques useful to determine the result of an esterification, the DS, and the pattern of functionalisation are described herein. The evaluation of the pattern of functionalisation is illustrated in detail for the most important polysaccharide ester, cellulose acetate. Detailed spectroscopic data for other polysaccharide esters are given in Chaps. 3 and 5.

From the synthesis chemist's perspective, the most reliable, powerful and efficient method for the detailed structure elucidation at the molecular level is NMR spectroscopy. A number of interesting new chromatographic tools have been developed over the last two decades, with a potential of gaining defined structural information, but they are combined with a variety of complex functionalisation steps making them susceptible to analytical errors and, therefore, should be used only by experienced analysts.

8.1 Chemical Characterisation – Standard Methods

Saponification of the ester function in the polysaccharide derivatives with aqueous NaOH or KOH and back-titration of the excess base is one of the oldest and easiest methods for the determination of the DS. In the case of short-chain aliphatic esters (C_2–C_4) of glucans, galactans and fructans, it is a fairly accurate method. For polysaccharide esters containing acidic (COOH) or basic (NH) functions, an absolute deviation may occur.

For longer aliphatic esters, strong hydrolysis conditions need to be applied (0.5 N NaOH, heating 48 h at 50 °C [372]). Saponification in solution, e.g. in acetone, has been used to determine the total acyl content in mixed polysaccharide esters, e.g. cellulose acetate propionates and acetate butyrates. This procedure overcomes some of the difficulties encountered in the commonly used heterogeneous saponification, in that it is independent of the condition of the samples, can be run in a shorter elapsed time, and is a little more accurate [373]. Analysis of polysaccharide esters containing many of the C_2–C_4 acid esters is carried out by completely hydrolysing the ester with aqueous alkali, treating with a high-boiling mineral acid in an amount equivalent to the alkali metal content, distilling the resulting mass to obtain all of the lower fatty acids in the form of a distillate, titrating a given amount of the distillate with standard alkali, and determining the different amounts of acids by titration after fractionated extraction [374].

Acid–base titration is an approach to evaluate ester cross-linking of polysaccharides with polycarboxylic acids such as 1,2,3,4-butanetetracarboxylic acid, by measuring the concentrations of ester and free carboxylic acid using calcium acetate back-titration [375, 376].

In the case of polysaccharide dicarboxylates and esters with hydroxy polycarboxylic acids, i.e. citric acid, malic acid or tartaric acid, potentiometric titration is used [377, 378]. Unsaturated esters such as starch acrylic acid esters have been characterised via the bromide/bromate method or with permanganometric titration [379].

For esters containing heteroatoms, a convenient method is elemental analysis. The DS is calculated by Eq. (8.1).

$$DS = \frac{\% \, \text{Analyte} \cdot M_r \, (\text{RU}) - 100 \cdot a \cdot M_r \, (\text{RU})}{100 \cdot b \cdot M_r \, (\text{Analyte}) \cdot M_r \, (\text{Introduced mass})} \qquad (8.1)$$

%Analyte (e.g. heteroatom) obtained by elemental analysis
a = number of analyte in the repeating unit
b = number of analyte in the introduced group
M_r (RU) = the molar mass of the repeating unit

8.2 Optical Spectroscopy

In addition to the characteristic signals for the polysaccharides given in Table 3.2, the strong C=O stretch band at 1740–1750 cm^{-1} is characteristic of an ester moiety.

For unsaturated esters, the signal is shifted to lower wave numbers (about 20 cm^{-1}) and, in esters with strong electron withdrawing groups, e.g. trifluoroacetates, is in the region of 1760–1790 cm^{-1} [188].

IR spectroscopy has been used for a quantitative evaluation of the amount of bound carboxylic acid and the distribution of the primary and secondary hydroxyl groups. It is a valuable tool for low-substituted derivatives and has been used to estimate the DS. Consequently, it is applied for the analysis of low-substituted starch acetates, showing good reproducibility [380]. Moreover, the FTIR bands assigned to sodium acetate produced by saponification of starch acetate can be detected by means of the ATR method, which can be used to determine the DS, showing a good correlation with values obtained from NMR [381].

In IR spectra of highly substituted cellulose acetates and mixed esters (0.15–3.0% solution in methylene chloride), two signals at 1752 and 1740 cm^{-1} are observed, corresponding to acyl moieties at positions 2 and 3 and at position 6 respectively (Fig. 8.2). The signal areas can be used for the calculation of the ratio of substitution at the different reactive sites.

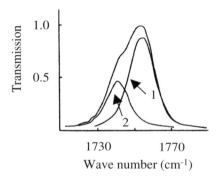

Fig. 8.2. The C=O signal region of an IR spectrum (0.15–3.0% solution in methylene chloride, measured in a KBr cell) of highly substituted cellulose acetates (1, C=O signal at positions 2 and 3, 2, C=O signal at position 6, adapted from [383])

The wave number is not influenced by the length of the chain in aliphatic acid esters, as shown in Fig. 8.3, and thus the technique can be extended to mixed esters.

Comparable information concerning the distribution of substituents is accessible by evaluation of the signal region of the OH groups (Fig. 8.4).

A signal at 3660 cm^{-1} corresponds to the primary OH unit. Furthermore, signals at 3520 and 3460 cm^{-1} are caused by secondary hydroxyl moieties. The absorption at 3580 cm^{-1} can be attributed to hydrogen bonding of the primary hydroxyl group [382, 383]. Line shape analysis or deconvolution of the spectra is helpful, and can be carried out with many modern FTIR instruments.

In addition to IR spectroscopy, Raman and NIR spectroscopy are frequently utilised for the investigation of polysaccharide esters. The potential of these two methods is discussed in [384]. DS determination for a number of cellulose derivatives, including tosyl cellulose and cellulose phthalate, has been carried out after calibration with standard samples of defined DS. It has been shown that confo-

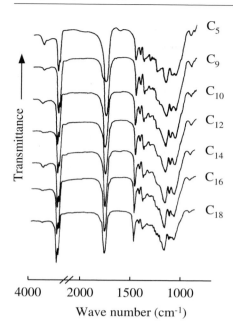

Fig. 8.3. FTIR spectra of aliphatic acid esters of cellulose with different numbers of carbons in the range C_5–C_{18} (reproduced with permission from [95], copyright Wiley VCH)

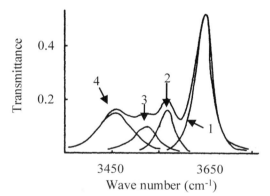

Fig. 8.4. IR spectrum (0.15–3.0% solution in methylene chloride, measured in a KBr cell) of cellulose acetates (OH group region, 1, primary OH, 2, hydrogen bond of primary OH group, 3 and 4, secondary OH functions, adapted from [382])

cal Raman spectroscopy is a very valuable method to study surface properties of such derivatives. Additionally, remote Raman sensing is a valuable tool for the determination of kinetic and chemical engineering data of esterification reactions, which are obtained in a direct and non-invasive inline manner by using remote Raman sensors. This is illustrated by the synthesis of cellulose acetate and cellulose phthalate. In Fig. 8.5, the development of Raman spectra versus time is shown for a reaction mixture consisting of cellulose dissolved in DMAc/LiCl and phthalic anhydride at 70 °C over 10 h. The disappearance of signals for the anhydride at 1760, 1800 and 1840 cm^{-1} is visible on the one hand. On the other hand, the development of a signal for the phthalate at 1720 cm^{-1} is observed.

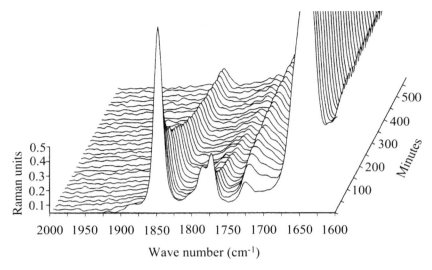

Fig. 8.5. The development of the Raman spectra versus time for a reaction mixture consisting of cellulose dissolved in DMAc/LiCl and phthalic anhydride at 70 °C over 10 h (reproduced with permission from [384], copyright ZELLCHEMING)

For the DS determination of maleinated starch by means of Raman spectroscopy, the calibration sets have very high linearity ($r > 0.99$). Combined with simple sample preparation, Raman spectroscopy is a convenient and safe method for the DS determination of polysaccharide esters [385].

UV/Vis spectroscopy is usually applied after ester hydrolysis. For calibration, standard mixtures of the polysaccharide and the acid are prepared and measurements at the absorption maximum of the acid are performed. DS determination is easily achieved by UV/Vis spectroscopy on mixtures of the saponified polysaccharide esters, e.g. for bile acid esters of dextran. The ester is dissolved in 60% aqueous acetic acid and a mixture of water/sulphuric acid (13/10, v/v) is added. The mixture is treated at 70 °C for 30 min and measured (after cooling) at 378 nm to obtain the amount of covalently bound bile acid [216, 217]. A similar technique is applied for the analysis of the phthaloyl content of cellulose phthalate after saponification [386] and for polycarboxylic acid esters, e.g. starch citrate [387].

An interesting and simple approach to determine the DS of starch esters with UV/Vis spectroscopy is the investigation of the iodine-starch ester complex. The effect of acetylation on the formation of the complex has been studied by monitoring the decrease in absorbance at 680 nm (blue value), which decreases by increasing the DS [381].

8.3 NMR Measurements

The application of NMR techniques was among the first attempts for the structure analysis of polysaccharide esters that exceeds the simple DS determination. The pioneering work of both Goodlett et al. in 1971 [388] using ^{1}H NMR spectroscopy, and of Kamide and Okajima in 1981 [389] applying ^{13}C NMR measurements on cellulose acetates opened major routes for further studies in this field, including complete signal assignment, the determination of the functionalisation pattern of polysaccharide esters depending on reaction conditions, and the establishment of structure–property relationships. Description of the sample preparation and representative signals of the polymer backbones are given in Sect. 3.2. NMR experiments on polysaccharide derivatives are most commonly performed in solution state. The solubility of the derivatives strongly depends on the DS and the type of polymer. Thus, a selection of NMR solvents used for the spectroscopy of polysaccharide esters is listed in Table 8.2. The preferred solvent for the investigation

Table 8.2. Typical NMR solvents used for the solution state ^{13}C- and ^{1}H NMR spectroscopy of polysaccharide esters

Name	^{1}H shift (multiplicity)	^{1}H shift of water	^{13}C shift (multiplicity)	Boiling point (°C)
Acetone-d_6	2.04 (5)	2.7	29.8 (7)	57
			260.0 (1)	
Acetonitrile-d_3	1.93 (5)	2.1	1.3 (7)	82
			118.2 (1)	
CDCl$_3$	7.24 (1)	1.5	77.0 (3)	62
D$_2$O	4.65 (1)			101.4
DMF-d_7	2.74 (5)	3.4	30.1 (7)	153
	2.91 (5)		35.2 (7)	
	8.01 (1)		162.7 (3)	
DMSO-d_6	2.49 (5)	3.4	39.5 (7)	189
CD$_2$Cl$_2$	5.32 (3)	1.4	53.8 (5)	40
Py-d_5	7.19 (1)	4.9	123.5 (5)	116
	7.55 (1)		135.5 (3)	
	8.71 (1)		149.9 (3)	
THF-d_8	1.73 (1)	2.4	25.3 (1)	66
	3.58 (1)		67.4 (5)	
Toluene-d_8	2.09 (5)	–	20.4 (7)	111
	6.98 (m)		125.2 (3)	
	7.00 (1)		128.0 (3)	
	7.09 (m)		128.9 (3)	
			137.5 (1)	
TFA-d_1	11.5 (1)	–	116.6 (4)	72
			164.2 (4)	

of partially substituted esters is DMSO-d_6, which dissolves the polymers within a wide range of DS values and is comparably inexpensive.

The application of ^{13}C NMR spectroscopy with focus on cellulose esters has been reviewed [63]. For the majority of $(1{\rightarrow}4)$ linked polysaccharides, e.g. starch and cellulose, esterification of the primary OH group results in a downfield shift (higher ppm values, between 2–8 ppm) of the signal for the adjacent carbon. In contrast, the signal of a glycosidic C-atom neighbouring carbon adjacent to an esterified OH moiety shows a high-field shift in the range of 1–4 ppm. The signal splitting and the corresponding shifts of the other carbon atoms of the polysaccharides strongly depend on the electronic structure of the ester moiety bound. This is illustrated for cellulose sulphuric acid half ester and cellulose acetate in Fig. 8.6. A complete assignment requires two-dimensional NMR techniques.

In the spectra of partially functionalised derivatives, a mix of spectra is observed (Fig. 8.6). In combination with the line broadening caused by different patterns of substitution, the spectra obtained are very complex. Nevertheless, the signal splitting and the intensities of the carbon atoms of the glycosidic linkage give an insight into the degree of functionalisation at the neighbouring positions.

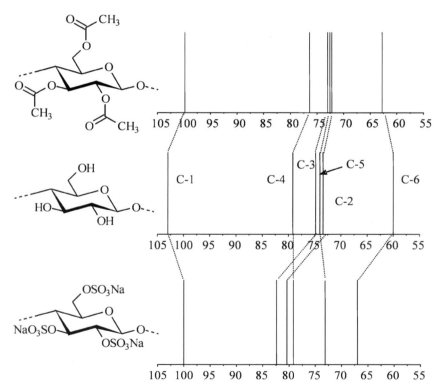

Fig. 8.6. Schematic ^{13}C NMR spectra of cellulose (*spectrum in the middle*) and completely sulphated (*lower picture*) as well as fully acetylated cellulose (*upper picture*) and the characteristic shifts caused by the esterification

The structure elucidation of organic esters of polysaccharides can be well illustrated with cellulose acetates, and the techniques can be similarly applied for other polysaccharide esters. Table 8.3 shows representative ^{13}C NMR spectroscopic data of cellulose triacetate.

Table 8.3. Chemical shifts of the ^{13}C NMR signals for cellulose triacetate (adapted from [390])

	δ (ppm)[a]	
	DMSO-d_6 90 °C	CDCl$_3$ 25 °C
C-1	99.8	100.4
C-2	72.2	71.7
C-3	72.9	72.5
C-4	76.4	76.0[b]
C-5	72.5	72.7
C-6	62.8	61.9

[a] Relative to CDCl$_3$ at 77.0 ppm or DMSO-d_6 at 35.9 ppm
[b] The coupled resonance overlaps with the solvent resonance

The assignment of the signals is based on the comparison of chemical shifts of model compounds such as peracetylated cellobiose, cellotetraose, cellopentaose and cellulose [391–393]. Data are available for the peracetylated homoglucanes pullulan [102] and dextran [58], i.e. the pullulan nonaacetate with its maltotriose structure of the polymer backbone (Table 8.4).

As described above, ^{13}C NMR spectra of cellulose acetates provide structural characterisation and determination of both total DS and the distribution of acetyl functions within the RU concerning positions 2, 3 and 6 in partially functionalised polymers [394, 395]. The investigation of cellulose acetate samples with DS values of 1.7, 2.4 and 2.9 reveals the same NOE for C-1–C-6, which confirms quantitative assessment of partial DS values at positions 2, 3 and 6 from the ^{13}C NMR spectrum. The signals at 59.0 ppm (C-6 unsubstituted), 62.0 ppm (C-6 substituted), 79.6 ppm (C-4, no substitution at C-3), 75.4 ppm (C-4 adjacent to substituted position 3), 101.9 ppm (C-1, no substitution at C-2) and 98.9 ppm (C-1 adjacent to substituted position 2) are used for the calculation.

The exact distribution of substituents in polysaccharide esters over a wide range of DS is not readily estimated by simple comparison of the relevant peak intensities. A major problem is the overlapping of signals at around 70–85 ppm, resulting from the unmodified C-2–C-5 and the corresponding acylated positions 2 and 3 as well as the influence of an acylated position 3 on the chemical shift of C-4. In addition, line broadening of the signals due to the ring carbons is frequently observed in the quantitative mode of ^{13}C NMR measurements. The fairly long pulse repetition time applied causes T_2 relaxation of the relevant signals.

Table 8.4. Assignment of the carbon signals in pullulan peracetate in ppm (position, see Fig. 2.4, adapted from [102])

Position	Chemical shift (ppm)	
	Maltotriose	C=O
A1	95.62	–
A2	70.58	170.72
A3	71.89	169.64
A4	72.76	–
A5	69.00	–
A6	62.84	170.32
B1	95.99	–
B2	71.32	170.45
B3	72.30	169.42
B4	73.88	–
B5	69.77	–
B6	63.10	170.34
C1	95.60	–
C2	70.10	170.52
C3	69.79	169.85
C4	68.39	169.02
C5	68.03	–
C6	64.80	–

Several authors have concentrated on assigning the signals of the C=O of the acetyl moieties [396]. Acetyl methyl and C=O signals appear as overlapped multipeaks, reflecting the detailed substitution pattern with regard to the 8 different RU as well as the hydrogen bond system of cellulose acetate with DS < 3. A preliminary assignment of the C=O signals of cellulose acetate is achieved by applying a low-power selective decoupling method to the methyl carbon atoms of the acetyl groups [397], and can be carried out via C-H COSY spectra of cellulose triacetate (Table 8.5, [398]).

An elegant analysis of the structure elucidation of cellulose [1-^{13}C] acetates prepared in different ways with a wide range of DS values is described by Buchanan et al. [399]. A total of 16 carbonyl carbon resonances can be identified using a variety of NMR techniques, including INAPT spectroscopy. The assignment differs somewhat from that given in Table 8.5. In the case of C-2 and C-3 carbonyl carbon resonance, it is possible to assign these resonances to repeating units with a specific pattern of functionalisation.

Table 8.5. Peak assignment of the carbonyl region in the ^{13}C NMR spectrum of cellulose acetate (adapted from [397])

δ (ppm)	Carbon at position	Functionalised glucopyranose unit
170.04	6	6-Mono
170.00	6	2,6-Di
169.94	6	2,3,6-Tri
169.89	6	3,6-Di
169.83	6	3,6-Di
169.60	3	3-Mono
169.46	3	3,6-Di
169.41	3	3,6-Di
169.35	3	3,6-Di
169.22	3	2,3-Di
169.11	3 (2)	2,3,6-Tri (2,6-Di)
168.93	2	2-Mono- 2,6-Di
168.79	2	2,3,6-Tri- 2,3-Di
168.71	2	2,3-Di

Although experimentally quicker and easier for quantitative evaluation of structural features, ^{1}H NMR spectroscopy on partially functionalised polysaccharide esters is limited because of the complexity of the spectra resulting from the un-, mono-, di-, and trisubstituted RU with different combinations of the functionalised sites (Fig. 8.7).

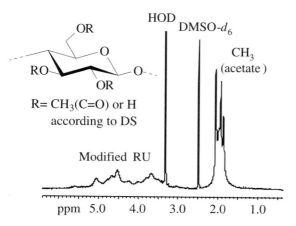

Fig. 8.7. ^{1}H NMR spectrum of a cellulose acetate (DS 2.37)

Nevertheless, cellulose acetate with a DS of 2.46 has been studied with phase-sensitive COSY and relayed COSY NMR spectroscopy. Comparison both of the spectra with simulated ones (nine different spectra) and model compounds, e.g. cellotetrose peracetate, shows that nine different types of spin systems are found. They are four types of 2,3,6-triacetyl glucose residues flanked by different acetyl glucose units, two different types of 2,3-diacetyl glucose residues, a 2,6-diacetyl glucose residue, and a 6-monoacetyl glucose residue [400].

In contrast to ^{1}H NMR spectra of partially acylated polysaccharides, completely substituted polymers yield assignable spectra, as shown for cellulose triacetate in Fig. 8.8 and Table 8.6.

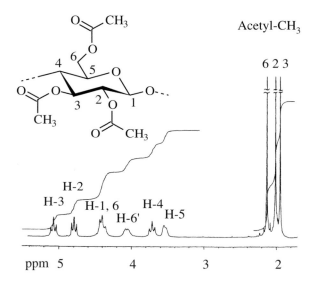

Fig. 8.8. ^{1}H NMR spectrum of a cellulose triacetate (reproduced with permission from [151], copyright Wiley VCH)

Table 8.6. Chemical shifts of ^{1}H NMR signals for cellulose triacetate

Signal	δ (ppm) of cellulose triacetate		
	DMSO-d_6 25 °C	DMSO-d_6 80 °C	CDCl$_3$ 25 °C
H-1	4.65	4.65	4.42
H-2	4.52	4.55	4.79
H-3	5.06	5.04	5.07
H-4	3.65	3.68	3.66
H-5	3.81	3.77	3.47
H-6′	4.22	4.26	–^{a}
H-6	3.98	4.04	4.06

^{a} H-6′ results from the signal splitting due to the neighbouring chiral C-5 and overlaps with H-1

A similar complete signal assignment and structural determination for other fully functionalised polysaccharide esters such as pullulan nonaacetate has been reported (Table 8.7).

Table 8.7. Chemical shifts of ^{1}H NMR signals for a pullulan nonaacetate (adapted from [102])

Position	Chemical shift (ppm)	
	Maltotriose	CH$_3$
A1	5.26	–
A2	4.71	2.02
A3	5.38	1.99
A4	3.88	–
A5	3.97	–
A6	4.17, 4.43	2.13
B1	5.10	–
B2	4.70	2.07
B3	5.49	1.96
B4	3.90	–
B5	3.92	–
B6	4.28, 4.41	2.14
C1	5.31	–
C2	4.76	2.01
C3	5.32	1.96
C4	5.13	2.05
C5	3.94	–
C6	3.57, 3.72	–

8.4 Subsequent Functionalisation

For a number of analytical techniques, complete subsequent derivatisation of the polysaccharide esters is necessary. In the case of ^{13}C NMR spectroscopy, this is applied to introduce moieties usable as internal probes, such as propionyl moieties. For ^{1}H NMR spectroscopy, complete functionalisation is an indispensable prerequisite for adequate spectral resolution leading to NMR data for quantitative evaluation. In order to utilise chromatographic tools for structure elucidation of polysaccharide esters, the transformation of the hydrolytically instable ester into a stable ether functionalisation is an essential step to obtain mixtures of RU with retained structural information, which can be separated via chromatography.

8.4.1 NMR Spectroscopy on Completely Functionalised Derivatives

To study the functionalisation pattern of partially derivatised polysaccharide esters, perpropionylation of the remaining hydroxyl groups is carried out and the

C=O carbons of the ester moieties are exploited as sensitive probe [401]. Complete propionylation is achieved by reaction of the polysaccharide ester with propionic anhydride, using DMAP as catalyst. The complete conversion of the hydroxyl groups is confirmed by ^{1}H NMR and IR spectroscopic analyses (no ν(OH) signal appears). Ester exchange reactions can be excluded by constant total acyl content at different reaction conditions and by propionylation experiments with polysaccharide triesters.

The range of C=O carbons in ^{13}C NMR spectra of a perpropionylated cellulose acetate (DS 1.43) is shown in Fig. 8.9. The signals appear clearly resolved and correspond to positions 2, 3 and 6 within the repeating unit. The triplet of the acetyl and the triplet of the propionyl moieties are distinctly separated from each other.

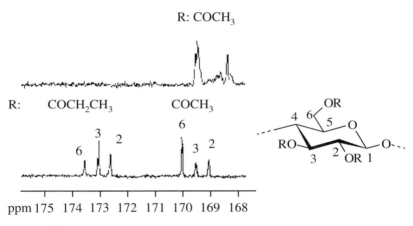

Fig. 8.9. ^{13}C NMR spectra of the carbonyl region of cellulose acetate (DS 1.43, *top*) and its propionylated product (*bottom*, reprinted from Carbohydr Res 273, Tezuka et al., Determination of substituent distribution in cellulose acetate by means of a ^{13}C NMR study on its propanoated derivative, pp 83–91, copyright (1995) with permission from Elsevier)

Quantitative-mode ^{13}C NMR measurements of perpropionylated samples give the partial DS at positions 2, 3 and 6. Typical ^{13}C NMR spectra of perpropionylated cellulose acetate samples with DS ranging from 1.0 to 2.4 are shown in Fig. 8.10. Quantitative-mode ^{13}C NMR spectroscopy (inverse gated decoupled experiments) is an expensive and time-consuming technique, as up to 20 000 scans are necessary for sufficient resolution.

^{1}H NMR spectroscopy for the structure determination of polysaccharide esters is much faster (usually only 16–64 scans) and less expensive. Standard ^{1}H NMR spectra are usable for precise quantification at a sufficient resolution. Consequently, ^{1}H NMR spectroscopy is being increasingly applied for the quantitative evaluation of the pattern of functionalisation of polysaccharide esters. It is most frequently utilised for structure determination after complete subsequent functionalisation of the unreacted OH groups of the polysaccharide derivative. The

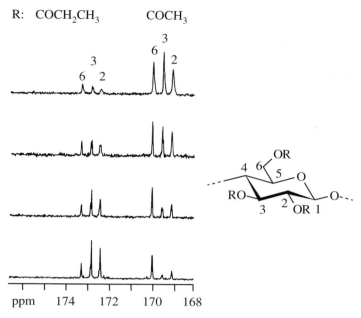

Fig. 8.10. Typical ^{13}C NMR spectra of perpropionylated cellulose acetate with DS 0.98, 1.43, 1.90 and 2.42 (from *top* to *bottom*, reprinted from Carbohydr Res 273, Tezuka et al., Determination of substituent distribution in cellulose acetate by means of a ^{13}C NMR study on its propanoated derivative, pp 83–91, copyright (1995) with permission from Elsevier)

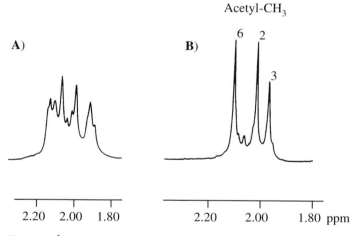

Fig. 8.11. ^{1}H NMR spectra of a cellulose acetate **A** before and **B** after deuteroacetylation (reproduced with permission from [151], copyright Wiley VCH)

complete modification results in an enormous simplification of the spectra, as shown in Fig. 8.11 for a cellulose acetate after deuteroacetylation. ^{1}H NMR spectroscopy after deuteroacetylation with acetyl-d_3-chloride or acetic anhydride-d_6

in Py was among the first attempts, and is still a common technique [151, 388]. The disadvantage is the rather expensive deuteroacetylation, which may be associated with a remarkable deviation in DS determination if the acetyl-d_3-chloride is contaminated with acetyl chloride. Nevertheless, it is a useful technique to determine partial DS values at positions 2, 3 and 6 in cellulose acetate, which can be readily calculated from the ratio of the spectral integrals of the protons of the RU and the methyl protons of the acetyl moiety.

Alternatively, trimethylsilylation of cellulose acetate with a wide range of DS values is carried out with N,O-bis(trimethylsilyl)acetamide and 1-methylimidazole in DMF at room temperature. Analysis of the samples by IR shows complete absence of hydroxyl groups. As evident in Table 8.8, the DS values obtained by integration of the O-acetyl and O-trimethylsilyl resonances are in excellent agreement with values provided by the suppliers and vide infra, with those found by chemical analysis [402]. Partial DS values are not accessible.

Table 8.8. Comparison of the DS for cellulose acetate obtained by silylation and NMR spectroscopy with values given by the suppliers. The DS values were calculated from the integrals of the ring hydrogen and O-acetyl resonances (method 1) or from the integrals of the O-trimethylsilyl and O-acetyl resonances (method 2) in the ^{1}H NMR spectra of the fully O-trimethylsilylated polymers (adapted from [402])

Sample[a]	DS		
	Reported	Method 1	Method 2
A	0.80	0.81	0.76
B	2.10	1.97	2.05
C	2.50	2.28	2.50
D	3.00	2.77	3.00
E	2.45	1.86	2.43

[a] Samples A–D are Eastman products, E is an Aldrich product

Perpropionylation can be used for the structure determination of a broad variety of polysaccharide esters. The determination of both the DS and partial DS can be achieved with the signals of the subsequently introduced ester moiety, making the method very reliable. This is shown for a commercial cellulose diacetate. Complete derivatisation of the OH groups is carried out by treatment with excess propionic anhydride (5 ml per 0.3 g ester) in Py (5 ml) for 16 h at 70 °C. Completeness of the reaction is confirmed by the disappearance of the ν(OH) band in the FTIR spectra. The mixed ester obtained is highly soluble in CDCl$_3$, and the standard ^{1}H NMR NMR spectrum (Fig. 8.12) shows the peaks for propionate functions at 1.03–1.07 ppm (CH$_3$ positions 2 and 3), 1.21 ppm (CH$_3$ position 6), and at 2.21–2.37 ppm for the CH$_2$ moiety. Three separate peaks are observed for CH$_3$ groups of acetyl moieties at 1.92 ppm (position 3), at 1.97 ppm (position 2), and at 2.08 ppm (position 6). The signals for protons of the AGU are found in the range 3.51–5.05 ppm. This agrees with results for regioselectively functionalised cellulose esters where full assignment of NMR signals is carried out by means of high-sensitivity HMBC techniques together with the conventional 2D-NMR techniques [403].

The AGU signals can be completely assigned, showing only one peak per proton (5.00, H-3, 4.73, H-2, 4.33, H-1,6, 3.99, H-6′, 3.64, H-4, 3.48 ppm, H-5), i.e. the existence of acetyl and propionyl moieties in the molecule does not induce a signal splitting of these protons. A ^{1}H,^{1}H-COSY NMR spectrum of this region is displayed in Fig. 8.12, confirming the assignment. Consequently, the spectral integrals of the protons of the AGU and the methyl protons of the propionyl moiety can be applied to calculate both the partial and the overall DS values of the acetate by means of Eqs. (8.2) and (8.3).

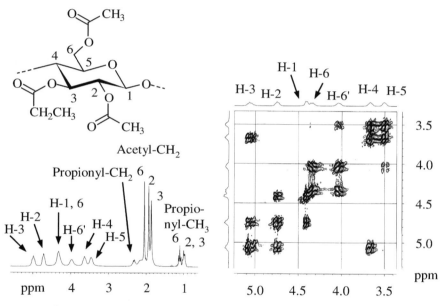

Fig. 8.12. ^{1}H NMR spectrum (*left*) and ^{1}H,^{1}H-COSY NMR spectrum (*right*, the area of the protons of the AGU is displayed) of cellulose acetate propionate prepared by complete propionylation of a commercial cellulose diacetate (CDCl₃, 32 scans)

$$DS_{Acyl} = 3 - \frac{7 \cdot I_{H,Propionyl}}{3 \cdot H_{H,AGU}} \tag{8.2}$$

$$DS_{Acyl}(n) = 1 - \frac{7 \cdot I_{H,Propionyl}(n)}{3 \cdot I_{H,AGU}} \tag{8.3}$$

I = integral

n = position 2, 3 or 6

Propionylation experiments of cellulose diacetate at temperatures between 60 and 120 °C show no significant changes in the DS and the distribution of the substituents. The results summarised in Table 8.9 confirm the accuracy of the method and yield a standard deviation of $S^2 = 1.32 \times 10^{-4}$. This method is reliable only after complete removal of all impurities, i.e. drying at 60 °C in vacuum. Water or free acetic acid give a high deviation.

Table 8.9. DS values calculated from ^{1}H NMR spectra of perpropionylated cellulose diacetate. The cellulose diacetate is propionylated twice (series 1 and series 2) and measured four times by ^{1}H NMR spectroscopy

DS series 1	DS series 2
2.35	2.37
2.35	2.37
2.32	2.38
2.32	2.38

DS values in the range 2.97–3.01 are found, which are within the standard deviation of the method (Table 8.10). Analysis of the partial DS_{Acetyl} is possible by evaluation of the propionate signal intensities, as described above, or from the three separate signals for CH_3 groups of the acetyl moieties at 1.92 ppm (position 3), at 1.97 ppm (position 2) and at 2.08 ppm (position 6).

Table 8.10. DS values of cellulose acetate with a broad variety of functionalisation determined via ^{1}H NMR spectroscopy after perpropionylation (adapted from [127])

No.	Molar ratio			DS_{Acetyl} at position			$DS_{Propionyl}$	Overall DS
	AGU	Acetyl chloride	Base	6	2, 3	Σ		
A1	1	1.0	1.2	0.35	0.13	0.48	2.49	2.97
A2	1	2.0	2.4	0.82	0.51	1.33	1.66	2.99
A3	1	3.0	3.6	0.91	0.65	1.56	1.41	2.97
A4	1	4.5	4.5	1.0	1.24	2.24	0.77	3.01
A5	1	5.0	6.0	1.0	1.62	2.62	0.37	2.99

Propionylation is helpful for the analysis of aromatic and unsaturated polysaccharide derivatives with ^{1}H NMR signals of the substituents in the region above 5.1 ppm. Starch-, dextran-, and cellulose furoates with additional peaks at 7.56, 7.20 and 6.50 ppm, and the 3-(2-furyl)-acrylic acid esters with additional peaks at 7.82, 7.50, 6.87, 6.57 and 6.23 ppm as well as alicyclic esters of cellulose, e.g. the ester of the adamantane carboxylic acid, have been analysed [168, 404, 405]. A representative ^{1}H,^{1}H-COSY NMR spectrum of a perpropionylated cellulose adamantane carboxylic acid ester is shown in Fig. 8.13. The signals for the protons of the two substituents are well resolved and can be used for the calculation of the DS according to Eqs. (8.2) and (8.3).

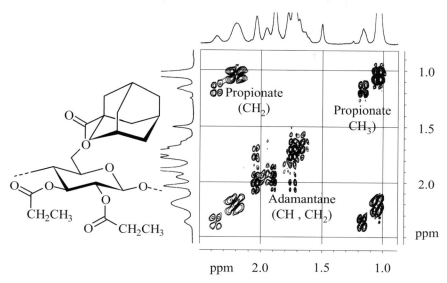

Fig. 8.13. ^{1}H,^{1}H-COSY NMR spectrum of a perpropionylated cellulose adamantane carboxylic acid ester. The area of the protons of the substituents is displayed (CDCl$_3$, number of scans 32)

In the case of polysaccharide esters with extended aliphatic moieties (signals in the range 0.8–3.2 ppm), such as fatty acid esters, the ^{1}H NMR spectra of the perpropionylated derivatives could not be assigned because of signal overlapping of the protons corresponding to the different ester moieties. Deuteroacetylation and, alternatively, per-4-nitrobenzoylation should be exploited. Nitrobenzoylation with 4-nitrobenzoyl chloride in Py gives signals only around 7.5–9.0 ppm in the ^{1}H NMR spectra.

Reactions of cellulose diacetate with 4-nitrobenzoyl chloride in DMF for 24 h at 60 °C confirm complete functionalisation of the free OH groups, and the product is soluble in CDCl$_3$. A standard ^{1}H NMR spectrum is shown in Fig. 8.14. Three separate peaks can be observed for the CH$_3$ group of the acetyl function at 1.88 ppm (position 3), at 2.02 ppm (position 2) and at 2.14 ppm (position 6). The signals for protons of the AGU between 3.46 and 5.09 ppm are equally well resolved as in the case of perpropionylated samples. No signal splitting of the AGU protons is induced by the existence of acetyl and 4-nitrobenzoyl moieties in the polymer. The DS$_{Acetate}$ can be determined from the ratio of the spectral integrals of the AGU protons and the methyl protons of the acetyl moiety. This is also useful for the DS determination of long-chain esters via the spectral integrals of the aromatic protons of the 4-nitrobenzoyl moieties. The content of acylation is calculated according to

$DS_{Acyl} = 3 - DS_{Nitrobenzoyl}$. The average value obtained by these two methods for a cellulose diacetate is 2.37, which is in good agreement with the values determined via perpropionylation and ^1H NMR spectroscopy.

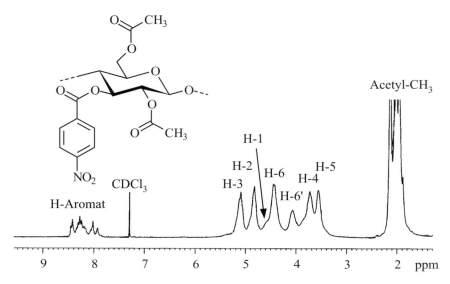

Fig. 8.14. ^1H NMR spectrum of a per-4-nitrobenzoylated cellulose acetate (CDCl$_3$, 32 scans)

A complex and ongoing problem is the determination of the distribution of the functional groups along the polymer chain and the supramolecular structure of acylated polysaccharides resulting from the functionalisation pattern both in the RU and along the chains. One possible approach is the application of NMR spectroscopy to samples with ^{13}C-labeled acetyl groups [399]. NOESY and ^{13}C NMR spectroscopy show that the ester forms a 5/4 helix. In addition, formation of domains can be investigated, which may give information on the pattern of substitution [406, 407]. NMR spectroscopy after enzymatic treatments of the cellulose esters leads to information about the composition of the polymer with regard to the eight different RU [408]. The sequence distribution of substituted glucopyranose units along the chain of cellulose acetate with a DS of 0.64 has been reconstructed with this technique, and is displayed in Fig. 8.15 [398].

8.4.2 Chromatographic Techniques

An important alternative to NMR spectroscopy is the determination of the inverse substitution pattern of the hydrolytically unstable polysaccharide ester by means of chromatographic techniques after subsequent functionalisation and depolymerisation, illustrated schematically in Fig. 8.16.

The chromatographic technique applied most frequently is GLC, usually in combination with MS. The advantage is that peaks can be unambiguously assigned

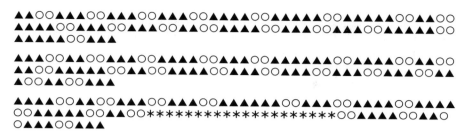

▲▲○○▲▲▲○○▲▲▲○○▲▲▲○○▲▲▲○○▲▲▲▲○○▲▲▲▲▲○○▲▲▲▲▲○○▲▲○○
▲▲▲▲○○▲▲▲▲○○▲▲▲○○▲▲○○▲▲▲▲○○▲▲▲○○▲▲▲○○▲▲▲▲▲○○
▲▲▲▲▲○○▲▲▲

▲▲▲○○▲▲○○▲▲▲○○▲▲▲▲○○▲▲▲○○▲▲▲▲▲○○▲▲▲▲○○▲▲○○
▲▲○○▲▲▲▲▲○○▲▲○○▲▲▲▲○○▲▲▲○○▲▲▲○○▲▲▲○○▲▲▲○○▲▲
▲○○▲▲○○▲▲▲

▲▲▲▲○○▲▲○○▲▲▲○○▲▲▲○○▲▲▲▲▲○○▲▲▲○○▲▲▲▲○○▲▲▲▲
○○▲▲▲▲▲○○▲▲○○∗∗∗∗∗∗∗∗∗∗∗∗∗∗∗∗∗∗∗∗∗∗∗∗○○▲▲▲▲○○▲▲○
○▲▲▲○○▲▲▲

Fig. 8.15. Sequence distribution of differently substituted glucose units along the polymer chain of a water-soluble CA with DS 0.64 and DP 96; ○ unsubstituted units; ▲ monosubstituted units; ∗ highly substituted regions (5 polymer chains are shown, adapted from [398])

Fig. 8.16. Analysis of polysaccharide esters with chromatographic methods after subsequent functionalisation

without the synthesis of standard compounds. This topic has been reviewed by Mischnick et al. [409]. HPLC has also been used as chromatographic tool, and the subsequent functionalisation is easier because the derivatives do not need to be volatile. Nevertheless, GLC still provides better resolution.

Subsequent functionalisation of polysaccharide esters may be carried out by etherification [410] and by carbamoylation [411, 412]. One of the first procedures was developed by Björndal et al. [413]. The derivatisation of cellulose acetate is performed by reaction with methyl vinyl ether in DMF in the presence of TosOH as catalyst. The anhydroalditol acetates are separated by means of gas chromatography after methylation according to Hakomori [81]. Results compared with data obtained by ^{13}C- and ^{1}H NMR spectroscopy show good agreement in the case of commercial cellulose monoacetate and cellulose diacetate (Table 8.11, [395]).

Table 8.11. Evaluation of the distribution of ester moieties in cellulose monoacetates (CMA) and cellulose diacetates (CDA) with chemical analysis using saponification and titration (Titration), with ^{13}C NMR spectroscopy (a) proton decoupled mode with NOE, (b) proton decoupled mode without NOE, ^{1}H NMR spectroscopy and GC after methylation and degradation (adapted from [395])

Sample	Method	DS of O-acetyl groups at position			Total DS
		2	3	6	
CMA	Titration	–	–	–	1.75
	^{13}C COM (a)	0.60	0.55	0.58	1.73
	^{13}C NME (b)	0.59	0.56	0.59	1.74
	^{1}H NMR	0.59	0.56	0.59	1.74
	GC	0.60	0.60	0.59	1.79
CDA	Titration	–	–	–	2.41
	^{13}C COM	0.84	0.83	0.72	2.39
	^{13}C NME	0.84	0.84	0.73	2.41
	^{1}H NMR	0.86	0.82	0.73	2.41
	GC	0.83	0.83	0.71	2.37

Methylation under neutral conditions with methyl triflate/2,6-tert-butylpyridine in trimethyl phosphate [414] or treatment with trimethyloxonium tetrafluoroborate in dichloromethane are more efficient [415]. These methylation procedures are used for derivatives with DS > 2. At lower DS, migration and chain degradation need to be considered. The methylated products are subjected to acyl–ethyl exchange and reductive cleavage [402]. The molar ratios of the eight differently substituted RU are determined by GLC (Fig. 8.17 and Table 8.12) for pure cellulose acetates.

In contrast to the pure cellulose acetates, the method is not capable of establishing the partial DS of each ester in mixed cellulose esters, e.g. cellulose acetate butyrates. For analysis of mixed esters, reductive treatment of the methylated samples with an excess of a Lewis acid in combination with triethylsilane is required. In detail, the methylated mixed cellulose esters are subjected to reductive cleavage

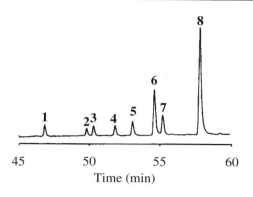

Fig. 8.17. GLC of anhydroalditol acetates derived from cellulose acetate with DS 2.50 (adapted from [402])

Table 8.12. Assignment of the eight differently substituted RU of anhydroalditol acetates derived from cellulose acetate with DS 2.50 (adapted from [402])

	R^1	R^2	R^3
1 (see Fig. 8.17)	CH_3	CH_3	CH_3
2	CH_3	C_2H_5	CH_3
3	CH_3	CH_3	C_2H_5
4	C_2H_5	CH_3	CH_3
5	CH_3	C_2H_5	C_2H_5
6	C_2H_5	C_2H_5	CH_3
7	C_2H_5	CH_3	C_2H_5
8	C_2H_5	C_2H_5	C_2H_5

at room temperature for 7 days in the presence of $(C_2H_5)_3SiH$ (35 mol/mol AGU), $CH_3SO_3Si(CH_3)_3$ (70 mol/mol AGU), and $BF_3{}^*O(C_2H_5)_2$ (17 mol/mol AGU). The procedure leads to a complete reduction of acyl residues to the corresponding alkyl ethers. GLC combined with CI-MS and EI-MS is carried out after acetylation of the samples with acetic anhydride/1-methylimidazole. Assignment of the 27 separate signals obtained on a Restek RTx-200 column is summarised in Table 8.13.

Structure analysis of polysaccharide sulphuric acid half esters is achieved via methylation under alkaline conditions. However, for trans-1,2-diol structures, as present in (1 → 4) linked glucans, an intramolecular nucleophilic displacement of the sulphate moieties under formation of an oxirane structure as intermediate may occur and needs to be avoided by optimised reaction conditions [416]. Due to the strongly enhanced acid lability of glycosyl linkages in 2-sulphates, reductive hydrolysis is applied to stabilise early-liberated glucose residues by direct reduction to glucitols [417, 418]. After complete hydrolysis, reduction and acetylation, par-

Table 8.13. Peak assignment (in the order of increasing retention times) of gas-liquid chromatograms of anhydroalditol acetates derived from cellulose acetate propionates by sequential permethylation, reductive cleavage, and acetylation (adapted from [402a])

R^2	R^3	R^6	Molecular mass $(g\,mol^{-1})$
CH_3	CH_3	CH_3	248
CH_3	C_2H_5	CH_3	262
CH_3	CH_3	C_2H_5	262
C_2H_5	CH_3	CH_3	262
CH_3	C_2H_5	C_2H_5	276
C_2H_5	C_2H_5	CH_3	276
C_2H_5	CH_3	C_2H_5	276
C_2H_5	C_2H_5	C_2H_5	290
CH_3	C_3H_7	CH_3	276
CH_3	CH_3	C_3H_7	276
C_3H_7	CH_3	CH_3	276
CH_3	C_3H_7	C_2H_5	290
CH_3	C_2H_5	C_3H_7	290
C_2H_5	C_3H_7	CH_3	290
C_3H_7	C_2H_5	CH_3	290
C_3H_7	CH_3	C_2H_5	290
C_2H_5	CH_3	C_3H_7	290
C_2H_5	C_3H_7	C_2H_5	304
C_2H_5	C_2H_5	C_3H_7	304
C_3H_7	C_2H_5	C_2H_5	304
CH_3	C_3H_7	C_3H_7	304
C_3H_7	C_3H_7	CH_3	304
C_3H_7	CH_3	C_3H_7	304
CH_3	C_3H_7	C_3H_7	318
C_3H_7	C_3H_7	C_2H_5	318
C_3H_7	C_2H_5	C_3H_7	318
C_3H_7	C_3H_7	C_3H_7	332

tially methylated glucitol acetates are obtained, which can be investigated by GLC. Other polysaccharide esters, including starch esters, e.g. acetates and benzoates, can be studied, too [419].

Methylation of polysaccharide esters, controlled depolymerisation and HPLC are applied alternatively to study their structure. The advantage is that aqueous solutions of the derivatised sugar units can be investigated, making the subsequent functionalisation rather easy. Permethylation with methyl triflate and subsequent hydrolytic depolymerisation with TFA can be used (analogues, Fig. 3.14, [188]). For separation of the methyl glucoses with inverse substitution patterns to the starting ester, a RP-18 column is applied. Assignment of the HPLC signals to the different patterns of substitution is feasible but quantification of the chromatograms is limited because of poor resolution. Nevertheless, the method can be used to determine the amount of different groups of substitution (un-, mono-, di-, and trisubstituted glucoses).

As described for NMR spectroscopic studies, a complex problem is the determination of the functionalisation patterns of polysaccharide esters along the polymer chain. One possible path consists in subsequent permethylation, deacetylative deuteromethylation under alkaline conditions, random cleavage, remethylation with methyl iodide-d_3, and FAB-MS analysis (Fig. 8.18). Comparison of the experimental data with those calculated for a random distribution of acetate groups gives information about the homogeneity of functionalisation along the chain [420]. Both undermethylation and migration, or cleavage of the primary functional groups during the methylation yield incorrect results.

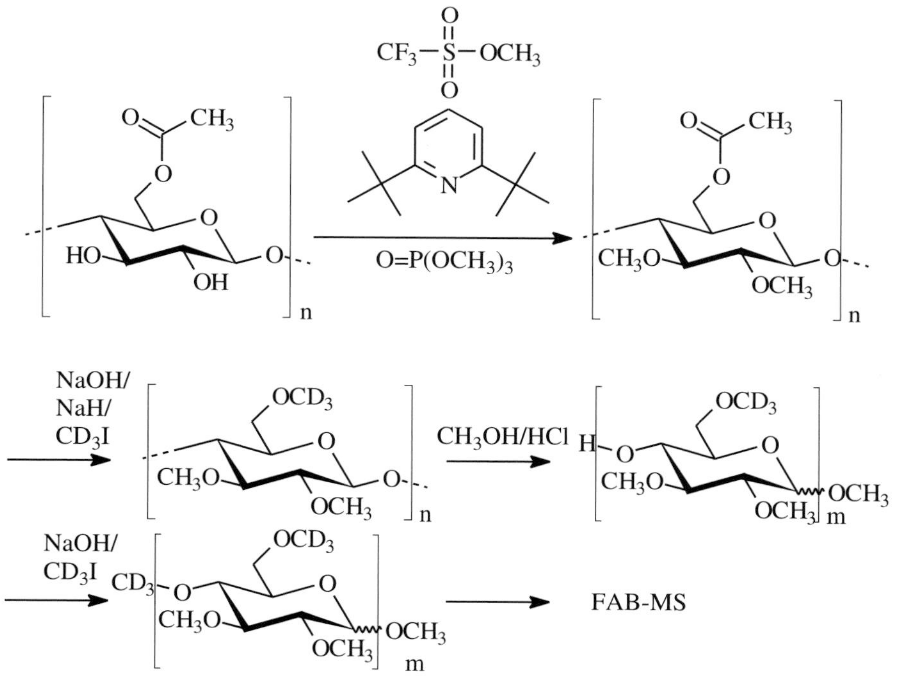

Fig. 8.18. Analytical path for the structure determination of cellulose acetate by FAB-MS after permethylation, perdeuteromethylation and random cleavage

Within the scope of this chapter, the analysis of the distribution of ester functions within the RU is broadly discussed. In the next chapter, synthetic paths for the achievement of defined functionalisation patterns in polysaccharide esters are described.

9 Polysaccharide Esters with Defined Functionalisation Pattern

Polysaccharide esters with a defined pattern of functionalisation are indispensable for the establishment of structure–property relations, e.g. for the solubility of cellulose acetate in function of the functionalisation pattern (Chap. 8). The defined functionalisation pattern may also yield unconventional thermal, optical and biological properties, as revealed for polysaccharide sulphuric acid half esters from dextran and curdlan with anti-HIV [421] and cancerostatic activity [422].

A number of approaches for the preparation of polysaccharide esters with a defined functionalisation pattern is known, applying mostly chemo- and regioselective synthesis and selective deacylation processes. Regioselective conversion may be realised by protective group techniques and so-called medium controlled reactions. However, the chemoselective functionalisation of polysaccharides has scarcely been exploited and is of special interest for the uronic acid-containing polymers, e.g. alginate, and for aminodeoxy polysaccharides (chitin and chitosan).

A selective esterification of the uronic acid units is discussed in Sect. 5.1.2. The polysaccharide is transferred into the acid form and then into the tetrabutylammonium salt, and finally this salt is converted homogeneously in DMSO with long-chain alkyl bromides (see Fig. 5.5, [5]).

In the case of chitin, the tailored modification is accomplished in different solvents (see Fig. 4.6). A number of valuable N-acetylated chitosan derivatives can be prepared in a mixture of methanol and acetic acid (Fig. 9.1, [423, 424]).

For polysaccharides containing exclusively hydroxyl groups, the modification reactions preferably occur at primary OH groups, especially if bulky carboxylic acid ester moieties are introduced.

A pronounced reactivity is observed for the OH group adjacent to the glycosidic linkage, due to electronic reasons. Consequently, for $(1\rightarrow4)$- and $(1\rightarrow3)$ linked polysaccharides, e.g. curdlan, starch and cellulose, the rate of esterification is usually in the order of position $6 > 2 > 3(4)$. For polysaccharides with no primary OH group, esterification at position 2 is the fastest conversion. Dextran shows an acylation reactivity of the OH moieties in the order $2 > 4 > 3$. Reaction paths leading to alternative patterns of esterification are described in the following sections.

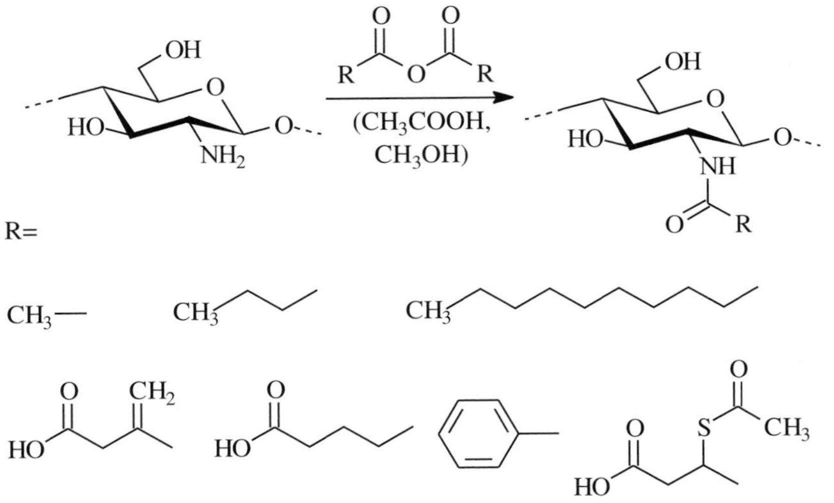

Fig. 9.1. N-acyl derivatives obtained by conversion of chitosan in acetic acid/methanol with carboxylic acid anhydrides

9.1 Selective Deacylation

Selective deacylation has been intensively studied for cellulose acetate. This is due to the fact that partially deacetylated cellulose acetates, e.g. cellulose diacetate, possess adjusted solubility (compare Table 8.1, Chap. 8) and can therefore be easily processed. The extent to which the polymer properties are controlled by the distribution of substituents within the RU is unknown. These properties may be additionally influenced by the distribution along the chain. Nevertheless, deesterification is used for the preparation of polysaccharide esters with unconventional functionalisation pattern within the RU. Polysaccharide acetates with adjusted functionalisation are valuable intermediates for the subsequent derivatisation, which leads (after adequate saponification) to subsequent derivatives with inverse functionalisation pattern. Cellulose triacetate is most commonly saponified directly (see Chap. 4). The hydrolysis is performed with aqueous H_2SO_4 and cleaves the primary hydroxyls that can later be reesterified [425, 426].

Different functionalisation patterns are obtained under different hydrolysis conditions [151]. For acidic hydrolysis of cellulose triacetate to products with DS values down to 2.2, the rate of deacetylation in position 6 and position 2 is comparable. If hydrolysis continues, deacetylation in position 2 is more pronounced, i.e. the acetyl functions in 6 are the most stable [89, 427]. Deacetylation in position 3 is the fastest (Fig. 9.2A). A different behaviour is observed if the hydrolysis with acetic acid/sulphuric acid is carried out directly after the complete functionalisation of cellulose. The rate of reaction is comparable for all three positions over the whole range of DS (Fig. 9.2B). Thus, cellulose acetate samples with an even distribution of substituents on the level of the AGU are obtained.

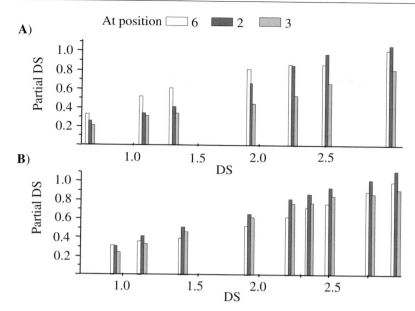

Fig. 9.2. Functionalisation pattern of differently prepared cellulose acetate samples as plot of the partial DS at positions 2, 3 and 6 versus the overall DS. Samples prepared via **A** acidic deesterification of cellulose triacetate and **B** acidic deesterification of cellulose triacetate directly after acetylation in *N*-ethylpyridinium chloride (reproduced with permission from [151], copyright Wiley VCH)

Even more pronounced is the preferred deacylation at the secondary positions for basic hydrolysis. Deacetylation of cellulose acetates in DMSO is achieved with hydrazine [89] or amines [360]. Adjustment of the 6 selectivity during the deesterification is feasible by deacetylation in the ternary mixture of DMSO/water/aliphatic amine (e.g. dimethylamine or hexamethylenediamine). Products with high DS at position 6, compared to the acetylation at positions 2 and 3, are obtained, as shown in Table 9.1 [360].

The specifically substituted cellulose acetate samples obtained are applied for the preparation of cellulose sulphuric acid half esters. The preferred functionalisation of the secondary OH groups shows a strong influence on the properties of the products, e.g. solubility, membrane formation, separation behaviour, and especially in interactions with human blood [360].

Cellulose acetate phosphates with defined functionalisation pattern have been prepared [428]. The mixed esters with phosphate moieties mainly in the positions 2 and 3 are manufactured by deacetylation of cellulose triacetates for 0.5–72 h at 20–100 °C with dimethylamine in aqueous DMSO. The deacetylation of cellulose acetate of DS 2.90 gives a product with DS 0.85 and partial DS_{Acetyl} of 0.05 (position 2), 0.15 (position 3) and 0.7 (position 6). Subsequent phosphorylation with polyphosphoric acid in DMF in the presence of tributyl amine yields cellulose acetate phosphate with DS_{Acetyl} 0.83 and DS_P 1.20, which may be deacetylated by

Table 9.1. Homogeneous deacetylation of cellulose triacetate in amine-containing media at 80 °C (adapted from [360])

Amine	Molar ratio		Time	DS at position		
	AGU	Amine	(h)	2	3	6
$NH_2-(CH_2)_6-NH_2$	1	2.3	2.5	0.80	0.80	1.00
$NH_2-(CH_2)_6-NH_2$	1	2.3	4.5	0.65	0.75	1.00
$NH_2-(CH_2)_6-NH_2$	1	2.3	9.0	0.45	0.55	0.90
$NH_2-(CH_2)_6-NH_2$	1	2.3	14.0	0.20	0.45	0.85
$NH_2-(CH_2)_6-NH_2$	1	2.3	24.0	0.05	0.10	0.60
$HN(CH_3)_2$	1	4.5	5.0	0.75	0.80	1.00
$HN(CH_3)_2$	1	4.5	11.0	0.50	0.50	1.00
$HN(CH_3)_2$	1	4.5	15.0	0.35	0.50	0.95
$HN(CH_3)_2$	1	4.5	20.0	0.30	0.40	0.90
$HN(CH_3)_2$	1	4.5	24.0	0.20	0.30	0.70

treatment with ethanolic NaOH to yield cellulose phosphate with DS_P 0.96 (partial DS_P of 0.77 at the secondary and of 0.19 at the primary positions).

A preferred esterification of the secondary hydroxyl groups is accomplished by conversions in derivatising solvents as well as via hydrolytically instable intermediates, as discussed in Sect. 5.1.4. The advantage of this approach is that isolation of the intermediate is not essential and the splitting of the intermediately formed function succeeds during workup. Cellulose sulphuric acid half esters with preferred functionalisation of positions 2 and 3 are accessible by reaction with SO_3-Py in N_2O_4/DMF homogeneously (with cellulose nitrite as an intermediate [192]) or by the conversion of cellulose trifluoroacetate [188, 429] and hydrolysis of the intermediate ester moiety.

To prepare regioselectively substituted cellulose acetate of low DS, purified acetyl esterases are used. Certain acetyl esterases cleave off the substituent from the 2 and 3 positions (carbohydrate esterase family 1 enzymes), whereas others deacetylate functional groups from position 2 (carbohydrate esterase family 5 enzymes) or from position 3 (carbohydrate esterase family 4 enzymes) [430, 431]. Regular deacetylation along the cellulose acetate chain is performed by the treatment of cellulose acetate (DS 0.9 and 1.2) with a pure *Aspergillus niger* acetyl esterase from the carbohydrate esterase family 1 [432]. Prior to the structure analysis, the enzymatically obtained fragments were separated by preparative SEC.

9.2 Protective Group Technique

The regioselective conversion of polysaccharides using protective group techniques is carried out with bulky ether functions such as triphenylmethyl- or silyl ethers selectively protecting the primary hydroxyl groups. The selective and direct protection of secondary hydroxyl groups is still an unsolved problem.

9.2.1 Tritylation

The bulky triphenylmethyl moiety is one of the oldest and cheapest protecting groups for primary hydroxyl moieties of polysaccharides. It is easily introduced by conversion of the polysaccharide suspended in Py with trityl chloride (3 mol/mol AGU) for 24–48 h at 80 °C. Most of the polymers dissolve during the reaction. An exception is chitin, in which complete tritylation of the primary position is limited and a DS of 0.75 is obtained only under rather drastic conditions (90 °C, 72 h, 10 mol reagent/mol RU, DMAP catalysis). Dissolution does not occur. Cellulose can be homogeneously tritylated in the solvent DMAc/LiCl. The polysaccharide trityl ethers are commonly soluble in DMSO, Py and DMF, and can be esterified without side reactions in these solvents. Deprotection is carried out with gaseous HCl in dichloromethane [398], aqueous HCl in THF [433] or preferably with hydrogen bromide/acetic acid [403].

The path is demonstrated by the synthesis of 2,3-di-O-acetyl-6-mono-O-propionyl cellulose (Fig. 9.3). The conversion of 6-O-triphenylmethyl cellulose with acetic anhydride in Py yields 2,3-di-O-acetyl-6-O-triphenylmethyl cellulose, which can be selectively detritylated with hydrogen bromide/acetic acid [403]. Subsequent acylation of the generated hydroxyl groups with propionic anhydride leads to a completely modified 2,3-di-O-acetyl-6-mono-O-propionyl cellulose. Starting with the propionylation, a product with an inverse pattern of functionalisation, i.e. 6-mono-O-acetyl-2,3-di-O-propionyl cellulose, is obtained, which is very useful for the assignment of peaks in the NMR spectra of cellulose esters [403].

Regioselectively substituted cellulose esters, e.g. propionate diacetate-, butanoate diacetate-, acetate dipropanoate-, acetate dibutanoate of cellulose, have been used to understand the thermal behaviour of mixed esters, compared with cellulose triester. DSC measurements have shown a correlation between the melting point and the length of the acyl groups at the secondary positions [434].

The regioselectively functionalised cellulose esters form crystals that can be studied by direct imaging of single crystals by atomic force microscopy (Fig. 9.4, [435]). The thickness is 29 nm for 2,3-di-O-acetyl-6-mono-O-propionyl cellulose and 45 nm for 6-mono-O-acetyl-2,3-di-O-propionyl cellulose. The dynamic structures formed in polar solvents of regioselectively substituted cellulose ester samples can be compared with those of commercial cellulose esters with random distribution, revealing large differences in the chain conformation, the solubility, and the clustering mechanism and structures [436, 437].

Protection of polysaccharides is very efficient with methoxy-substituted trityl moieties. It increases both the rate of conversion towards the protected polysaccharide and the rate of the deprotection step [433]. This is illustrated for the synthesis of protected cellulose in DMAc/LiCl (Table 9.2). In view of the pronounced selectivity, the stability of the protected cellulose, the selective detritylation and the price, protection with 4-monomethoxytrityl chloride is recommended. ^{13}C NMR spectroscopy is used to confirm the purity of the 4-monomethoxytrityl protected cellulose (Fig. 9.5). Complete detritylation of the protected polysaccharide is achieved with aqueous HCl in THF for 7 h.

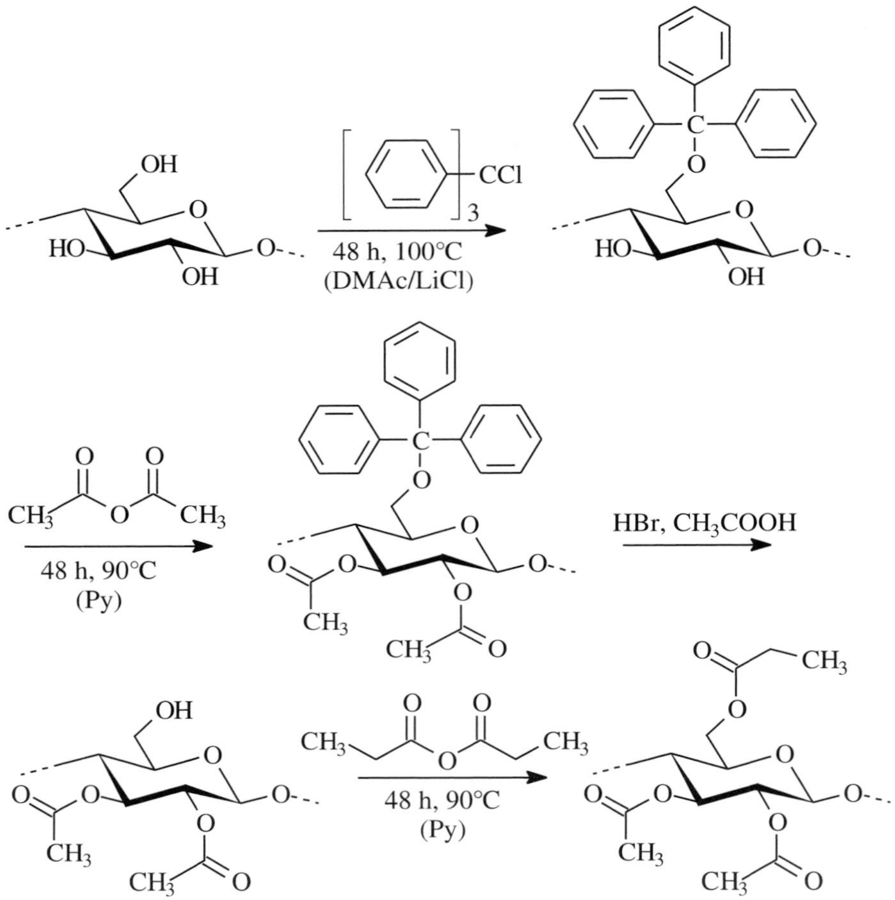

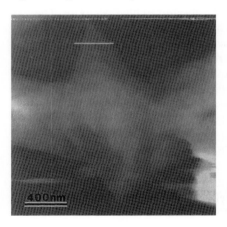

Fig. 9.3. Regioselective acylation of cellulose via 6-mono-*O*-trityl cellulose (adapted from [403])

Fig. 9.4. Atomic force microscopy image of single crystals of 2,3-di-*O*-acetyl-6-*O*-propanoyl cellulose (adapted from [435])

400nm

Table 9.2. Tritylation of cellulose with different trityl chlorides (3 mol/mol AGU, in DMAc/LiCl at 70 °C) and detritylation (37% HCl aq. in THF, 1:25 v/v, adapted from [438])

Substituent	Protection			Deprotection
	Time (h)	DS	Rate	Rate
Trityl	4	0.41	1	1
Trityl	24	0.92		
Trityl	48	1.05		
4-Monomethoxytrityl	4	0.96	2	18
4-Monomethoxytrityl	24	0.92		
4-Monomethoxytrityl	48	0.89		
4,4'-Dimethoxytrityl	4	0.97	2×10^5	100
4,4',4''-Trimethoxytrityl	4	0.96	6×10^6	590

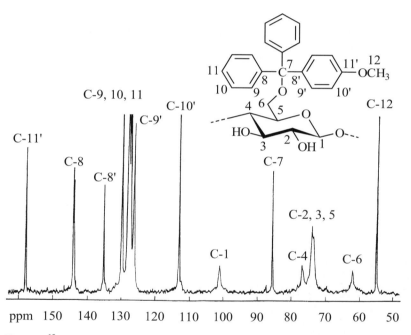

Fig. 9.5. ^{13}C NMR spectrum of 6-mono-*O*-(4-monomethoxy)trityl cellulose, DS 1.03 (reproduced with permission from [433], copyright Wiley VCH)

2,3-*O* functionalised cellulose sulphuric acid half esters are synthesised with a SO$_3$-Py- or SO$_3$-DMF complex (Table 9.3, [439]). This path can be applied for most polysaccharides with primary OH groups, including (1→3)-glucans such as curdlan [422,440]. Both DMAc/LiCl and DMSO are suitable solvents for the tritylation of starch but the highest DS of trityl groups obtained after a single conversion

Table 9.3. Regioselective cellulose sulphuric acid half esters prepared via 6-*O*-(4-mono-methoxy)triphenylmethyl cellulose (MMTC) and subsequent deprotection (adapted from [439])

MMTC	Reaction conditions		Product
DS	Agent	Time (h)	DS
0.98	SO$_3$-Py	2.0	0.30
0.98	SO$_3$-DMF	2.5	0.57
0.98	SO$_3$-Py	2.5	0.70
0.98	SO$_3$-Py	4.5	0.99

step was 0.77. A complete functionalisation of primary OH groups is achieved only with unsubstituted triphenylmethyl chloride. In the case of monomethoxy-triphenylmethyl chloride as reagent, an additional conversion step is necessary to synthesise products with DS 1. These procedures are less selective, compared with the single-step tritylation [441].

Moreover, regioselectivity can be achieved by enzymic transesterification, as shown for regenerated cellulose, 6-*O*-trityl cellulose and 2,3-*O*-methyl cellulose, when reacted with vinyl acrylate under enzymic catalysis (subtilisin Carlsberg). When the OH group at position 6 is blocked, enzyme-catalysed transesterification is not observed – even the OH moieties at positions 2 and 3 are free [442].

9.2.2 Bulky Organosilyl Groups

The protection of the primary hydroxyl groups in polysaccharides, and hence the preparation of mixed polysaccharide derivatives regioselectively esterified at the secondary positions is based on the introduction of TBDMS- and TDMS moieties. The selective protection of starch dissolved in DMSO is carried out with a mixture of

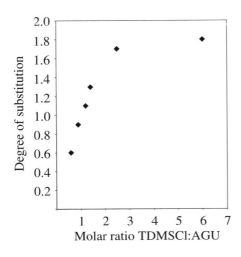

Fig. 9.6. DS of TDMS starch in function of the amount of TDMSCl during silylation in DMSO/Py (adapted from [443])

TDMSCl/Py (1.2 mol/mol AGU) for 40 h at 20 °C. The utilisation of higher amounts of silylating reagents leads to derivatives with DS up to 1.8 (Fig. 9.6). Subsequent homogeneous acetylation can be carried out in THF with acetic anhydride/Py [443].

Protection of cellulose in DMAc/LiCl has been reported with both TDMSCl/Py and TBDMSCl/Py (Table 9.4, [317, 444]). In the case of protection with TDMS moieties, a remarkable difference in selectivity is observed depending on the reaction conditions, which can be used for controlled derivatisation (Fig. 9.7).

6-O-TDMS cellulose carrying 96% of the silyl functions in position 6 is obtained by heterogeneous phase reaction with TDMSCl in the presence of ammonia-

Table 9.4. Silylation of cellulose with TBDMSCl and TDMSCl in DMAc/LiCl (5% cellulose, 8% LiCl, 1.1 mol Py/mol chlorosilane, adapted from [444])

| Molar ratio | | | DS | Solubility | | |
AGU	TBDMSCl	TDMS-Cl		DMF	THF	CHCl$_3$
1	1.0	–	0.67	+	–	–
1	1.5	–	0.96	+	–	–
1	3.0	–	1.53	–	+	+
1	–	1.0	0.71	+	–	–
1	–	1.5	0.92	+	–	–
1	–	3.0	1.43	–	+	+

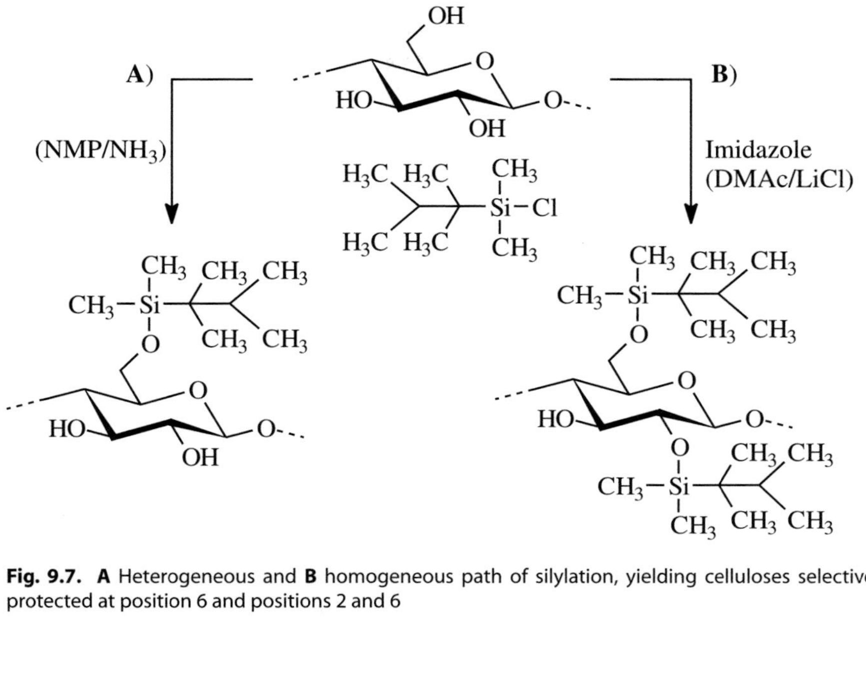

Fig. 9.7. **A** Heterogeneous and **B** homogeneous path of silylation, yielding celluloses selectively protected at position 6 and positions 2 and 6

saturated polar-aprotic solvents, e.g. NMP at −15 °C. In contrast, the homogeneous conversion of cellulose in DMAc/LiCl with TDMSCl in the presence of imidazole yields a 2,6-*O*-TDMS cellulose. Thus, selective protection of position 6 or the selective protection of positions 6 and 2 can be achieved. Acetylation is feasible either with acetyl chloride in the presence of a tertiary amine such as TEA [428] or with acetic anhydride/Py yielding the peracylated products. The selectivity of the conversion is illustrated by means of ^{1}H NMR and ^{1}H,^{1}H-COSY NMR spectroscopy (Fig. 9.8), studying 2,3-di-*O*-acetyl-6-mono-*O*-TDMS cellulose [445].

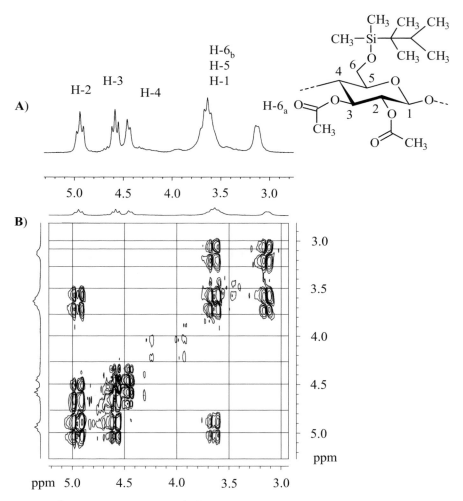

Fig. 9.8. ^{1}H NMR spectrum (**A**) and ^{1}H,^{1}H-COSY NMR spectrum (**B**) of 2,3-di-*O*-acetyl-6-mono-*O*-TDMS cellulose (reprinted from Cellulose 10, Silylation of cellulose and starch – selectivity, structure analysis, and subsequent reactions, pp 251–269, copyright (2003) with permission from Springer)

Derivatisation, desilylation and subsequent acetylation can be applied for the preparation of 6-mono-*O*-acetyl-2,3-di-*O*-methyl cellulose, 3-mono-*O*-methyl-2,6-di-*O*-acetyl cellulose or 3-mono-*O*-allyl-2,6-di-*O*-acetyl cellulose [446], confirmed by ^{1}H,^{1}H- and ^{1}H,^{13}C-COSY NMR spectra (Fig. 9.9).

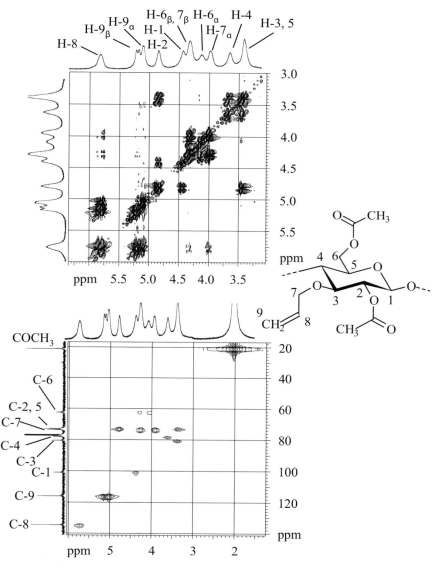

Fig. 9.9. ^{1}H,^{1}H- (*top*) and ^{1}H,^{13}C-COSY NMR spectrum (*bottom*) of 3-*O*-allyl-2,6-di-*O*-acetyl cellulose (reproduced with permission from [446], copyright Wiley VCH)

Deprotection is carried out by treatment of the polysaccharide derivatives with TBAF in THF solution. This is suitable for the preparation of regioselectively functionalised mixed ether esters. However, selective deprotection of silylated polysaccharide esters is not possible; treatment of the derivatives with TBAF results in the removal of both the protective group and the ester moiety.

9.3 Medium Controlled Selectivity

Starch can be selectively functionalised in position 2. It is assumed that the sterically preferred conversion of the primary position is hindered by interaction of this moiety with solvent molecules. Therefore, this level of selectivity is called "medium controlled". Selectivity is found for the transesterification of starch with vinyl esters of carboxylic acids in the presence of a catalyst (see Sect. 5.3.1, [245]). A broad variety of salts are able to catalyse this reaction (Table 9.5). Comparable syntheses have been accomplished with long-chain aliphatic acid vinyl esters, e.g. vinyl laurates, and with aromatic derivatives such as vinyl benzoates (Table 9.5). In contrast, the conversion with an anhydride does not show a pronounced selectivity.

Table 9.5. DS and partial DS at positions 2, 3 and 6 after acetylation of starch (Hylon VII) in DMSO with 2% (w/w) of different catalyst and acetylating agents (40 °C, 70 h, 2.3 mol acetylating agent/mol AGU, adapted from [245])

Reaction conditions		Starch acetate			
Catalyst	Acylating agent	DS	Partial DS at position		
			2	3	6
Na_2HPO_4	Vinyl acetate	1.00	1.00	0	0
Sodium citrate	Vinyl acetate	0.80	0.80	0	0
Na_2CO_3	Vinyl acetate	1.00	1.00	0	0
K_3PO_4	Vinyl acetate	1.03	1.00	0.03	0
Magnesium acetate	Vinyl acetate	1.15	1.00	0.10	0.05
Sodium acetate	Vinyl acetate	1.82	0.91	0.54	0.37

Even more astonishing is the selective tosylation of position 2 of starch in DMAc/LiCl, as discussed in Sect. 6.2. A significant influence of the reaction media appears also by silylation reactions of cellulose (cf. Fig. 9.7).

Tailored functionalisation patterns obtainable with the synthesis methods described above lead to structure–property relations and thereby to a variety of novel applications, as discussed in the following chapter.

10 Selected Examples of New Applications

Over 76 000 references regarding the application of polysaccharide esters are referenced in Scifinder® (American Chemical Society, July 2005). The most numerous products are cellulose esters (about 57 500 references), compared to starch- (about 9500 references), dextran- (about 7900 references), chitin- (about 430 references) and curdlan esters (about 130 references). This pronounced importance of cellulose esters and the wide scope of "industrial applications" are illustrated in Table 10.1.

Table 10.1. Some major applications of organic cellulose esters and the amounts produced in 1985 (adapted from [94])

Cellulose ester	DP	DS		Principal application	Amount (t/year)
		Acetyl	Propyl or butyl		
Triacetate	150–360	2.8–3.0	–	Textile fibres	280 000
				Photo film, foils, insulating coatings	60 000
Diacetate	100–200	2.5	–	Filter tow	370 000
				Thermoplastic mass	
		2.4	–	Viscose silk, foils	
		2.4	–	Thermoplastic mass	60 000
		2.3	–		
Acetopropionate	150–200	0.3	2.3	Thermoplastic mass	
Acetobutyrate	100–150	2.1	0.6	Raw materials for coating and insulation, foils	5000
		2.0	0.7	Foils, films	
		1.0	1.6	Thermoplastic mass	40 000
		0.5	2.3	Melt dipping mass	
				Σ	815 000

Among recent developments are polysaccharide esters for modern coatings, controlled release applications, biodegradable polymers, composites, optical film applications, and membranes. The tailored modification of properties can be accomplished by multiple esterification, i.e. two or even more different ester moieties are introduced where the one determines properties necessary for processing and the other ester group induces a specific product feature. This approach, for a huge variety of modified polysaccharides with focus on cellulose esters, has been excellently reviewed [447]. A selection of typical cellulose esters is given in Fig. 10.1.

The trends in polysaccharide ester utilisation are discussed by Glasser in [448], evaluating the amount of recent publications in the field of cellulose esters. The type of journals publishing the largest numbers of work recently (Table 10.2) indicates the increasing scientific interest in exploiting the high tendency of polysaccharide esters towards the formation of superstructures, the biological activity, and the biocompatibility resulting in the development of new separation techniques, biomedical devices and pharmaceuticals. Some selected developments will be discussed to illustrate the potential of new polysaccharide esters.

Table 10.2. Journals currently (last 3 years) publishing most frequently in English on cellulose esters (reproduced with permission from [448], copyright Wiley VCH)

Journal name	Number of publications (last 3 years)
Journal of Membrane Science	27
Journal of Applied Polymer Science	24
Drug Development and Industrial Pharmacy	12
Biomaterials	10
Polymer	10
Cellulose	9
Journal of Controlled Release	9
International Journal of Pharmaceutics	8

10.1 Materials for Selective Separation

Polysaccharide derivatives are well established as membrane materials and as selective stationary phases in chromatography. For a comprehensive overview of cellulosic materials for ultrafiltration, reversed osmosis, and dialysis, see Refs. [447,449]. The defined superstructure of polysaccharides and polysaccharide derivatives seems to be responsible for their high efficiency in separation processes, particularly for chiral resolution. The specific interaction of chiral molecules is observed with the pure polysaccharides and their derivatives leading to similar chiral discrimination. The reason could be a comparable superstructure of the polysaccharide and the fully functionalised ester, as has been concluded for curdlan and curdlan triacetate. The unit cell contains six RU related by 6/1-helical symmetry, which is essentially the same as the backbone conformation of one of the modifications (form I) of curdlan [450]. The chiral separation applying polysaccharide esters is not caused by hydrogen-bonding interaction with the solute, as determined for other chiral phases [451]. In contrast, the conformational regularity

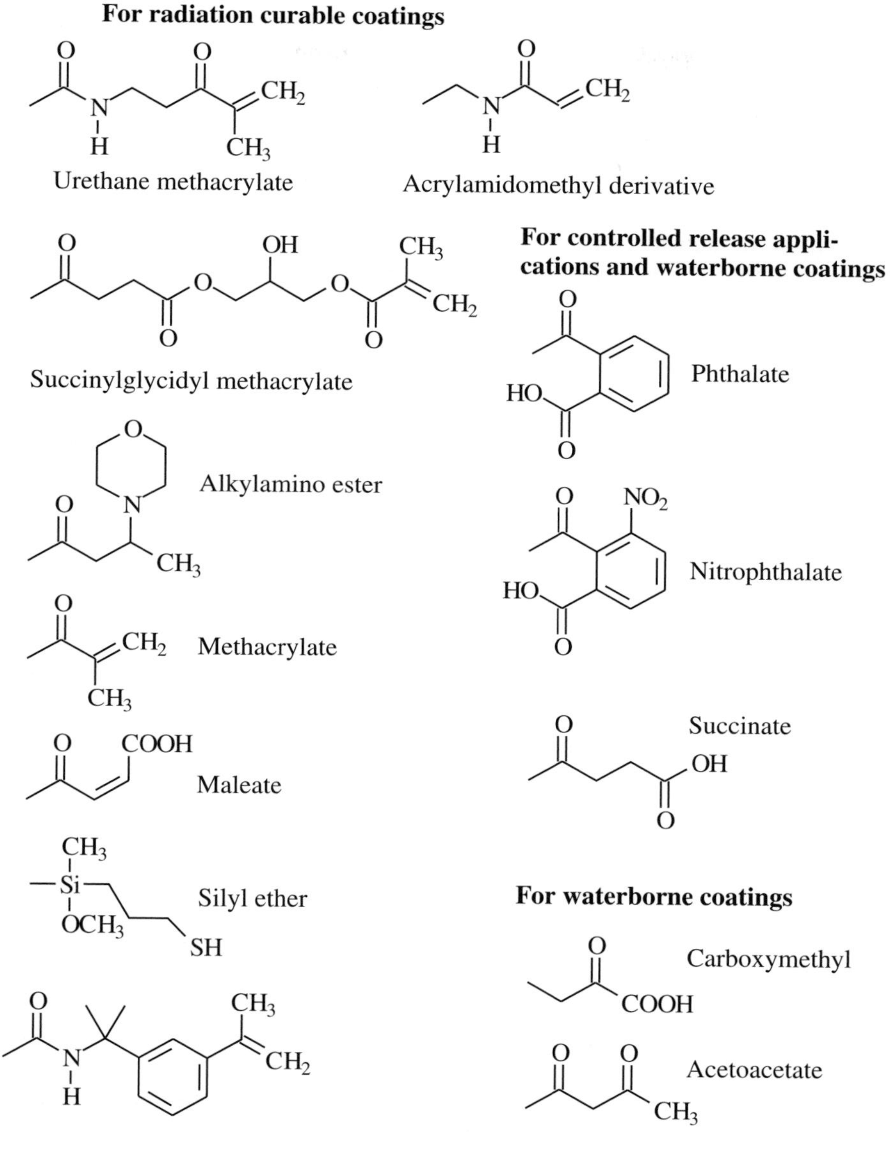

For radiation curable coatings

Urethane methacrylate

Acrylamidomethyl derivative

Succinylglycidyl methacrylate

For controlled release applications and waterborne coatings

Phthalate

Alkylamino ester

Nitrophthalate

Methacrylate

Maleate

Succinate

Silyl ether

For waterborne coatings

Carboxymethyl

Acetoacetate

m-Isopropenyl-1,1'-dimethylbenzylcarbamate

Fig. 10.1. Selection of cellulose esters for modern coatings and controlled release applications (adapted from [447])

achieved by different solvent treatment of cellulose triacetate during solidification on column packing materials exhibits a remarkable effect on the chiral separation properties, leading to the conclusion that the superstructure has a crucial influence [452].

10.1.1 Stationary Phases for Chromatography

In addition to the widely used substituted phenylcarbamates of polysaccharides, which are utilised after covalent immobilisation on the surface of silica gel [453], the triesters of polysaccharides [447] are exploited as stationary phases for chromatography (Table 10.3).

Table 10.3. Examples of new cellulose esters as stationary phases in chromatography

Cellulose ester	Remarks	Ref.
Acetate	Discrimination of racemic thiosulphinate	[454]
Benzoate	Discrimination of R(+)- and S(−)-benzyl-3-tetrahydrofuroates	[455]
4-Methyl-phenylbenzoat	Discrimination of thiazolo benzimidazoles	[456]
Alkenoxybenzoyl/benzoate 10-Undecenoyl/benzoate 10-Undecenoate/4-methyl-benzoate	For covalent binding to a silica surface via radical grafting with vinyl or allyl silica	[457], [458]
Acrylates, methacrylate	For covalent binding to a silica surface	[459]

Interestingly, stationary phases based on regioselectively substituted cellulose esters show chiral discrimination dependent on the distribution of the ester moieties, as demonstrated for 6-O-acetyl-2,3-di-O-benzoyl cellulose and 2,3-di-O-acetyl-6-O-benzoyl cellulose [460].

10.1.2 Selective Membranes

The most common polysaccharide membranes are based on cellulose esters, which have found applications in all fields of separation processes [447]. In addition to cellulose, galactomannan- and curdlan esters are potential film-forming materials. Thus, guar gum formate can be processed into an ultrathin semipermeable membrane [181], and curdlan acetate provides ultrafiltration membranes by casting from solutions containing HCOOH and additives such as water and DMF [461].

New paths towards selective films include the defined establishment of supramolecular structures by electrostatic interactions, and the concept of molecular imprinting. Polyelectrolyte complexes are used for sulphated polysaccharides, e.g. from sulphuric acid half esters of cellulose and poly(diallyldimethylammonium chloride) [462]. These polyelectrolyte complexes can be used for the formation of capsules with a defined cut-off value for the immobilisation of biological matter, e.g. yeast [463, 464]. Capsules based on cellulose sulphuric acid half esters are applied in an in situ chemotherapy strategy with genetically modified cells in an immuno-protected environment, and may prove useful for solid tumour therapy in man [465, 466].

Loading of a surface with ionic functions is carried out yielding compounds with a pronounced biological selectivity. By treatment of a porous NH_3^+-containing polypropylene membrane with the sulphuric acid half ester of dextran, a material for the convenient removal of the human immunodeficiency virus (HIV) and related substances from blood, plasma or other body fluids has been developed. Filtration of HIV-containing human plasma results in 99.2% removal of HIV [467].

Molecularly imprinted polymeric membranes are prepared from cellulose acetate with benzyloxycarbonyl-L-glutamic acid and carbobenzoxy-D-glutamic acid. The imprinted polymeric material recognises the L-glutamic acid in preference to D-glutamic acid, and vice versa, from racemic mixtures [468, 469]. Polymer blends containing cellulose acetate and sulphonated polysulphone are a matrix for the preparation of molecularly imprinted materials via phase inversion from a casting solution containing a template. Membranes are obtained with Rhodamine B as template. Results for rebinding of Rhodamine B during filtration through the imprinted as well as blank membranes, prepared without Rhodamine B, provide evidence for surface imprinting (Fig. 10.2, [470]).

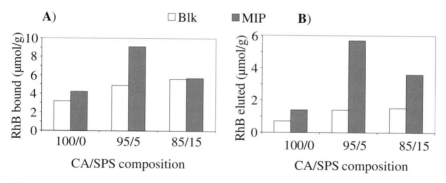

Fig. 10.2. Results from membrane solid-phase extraction for molecularly imprinted (MIP) and blank (Blk) membranes at varied cellulose acetate (CA):sulphonated polysulphone (SPS) ratio. **A** Bound Rhodamine B (Rh B) after filtration of 10 ml 10^{-5} M solution in water, and **B** Rh B eluted with 10 ml methanol after binding and subsequent washing with water and 2 M NaCl solution (membrane area 3.5 cm^{-2}, thickness $\sim$ 150 µm, adapted from [471], reproduced by permission of The Royal Society of Chemistry)

The surfaces can be studied in detail with scanning force microscopy and the gas adsorption isotherm method. Significant differences in pore structure between imprinted and blank membranes are found, which clearly correlate with the imprinting efficiency (Fig. 10.3, [471]).

A)   B)

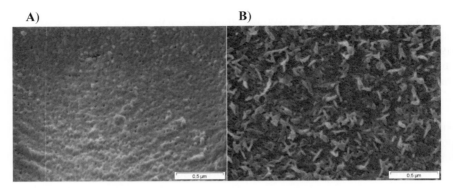

Fig. 10.3. High-resolution SEM micrographs of **A** imprinted and **B** blank membranes (cellulose acetate:sulphonated polysulphone 95:5) showing significant differences in the structure of the top layer (adapted from [471], reproduced by permission of The Royal Society of Chemistry)

10.2 Biological Activity

The most remarkable biological activity is observed for sulphated polysaccharides; naturally occurring polysaccharides play an important role in various biological systems, such as heparine in anticoagulant processes. A comparable behaviour is observed for semisynthetic polysaccharide sulphuric acid half esters and has been studied intensively because heparin is the drug of choice in clinical presurgical and postsurgical prophylaxis of thrombotic events. However, heparine exhibits a number of side effects caused by its chemical inhomogeneity and the variability of its physiological activities [472]. A summary of bioactive polysaccharide esters is given in Table 10.4.

Investigations of sulphuric acid half esters of cellulose and dextran suggest that the anticoagulant activities of these compounds are at least partially mediated through antithrombin III [476]. The anticoagulant activity is influenced by the pattern of functionalisation. For the cellulose ester, the sulphation of the secondary OH groups is a predominant factor in the anticoagulant activity, and the molecular mass is only of minor importance. In contrast, the toxicity is influenced both by the substituent distribution and the molecular mass [488].

Sulphated homopolysaccharides such as dextran and cellulose esters show potent virucidal activity against human T-cell lymphotropic virus type III (HTLV-III). In contrast, neutral homopolysaccharides have no effects and sulphated heteropolysaccharides exhibit only little effect on HTLV-III activities. This suggests that the sulphate group and the type of polysaccharide are most important in inhibiting growth of HTLV-III [489]. A new concept is the covalent binding of anti-HIV agents such as azidothymidine on sulphated curdlan with ester bonds, yielding a polymer with anti-HIV activity both via the controlled release of the azidothymidine from the polysaccharide ester carrier by enzymatic hydrolysis in living organs and via the polysaccharide ester structure itself [490]. To release

Table 10.4. Examples of polysaccharide esters with pronounced biological activity

Polysaccharide sulphuric acid half esters of:	Biological activity	Remarks	Ref.
Curdlan	Anticoagulant properties	Inhibitory activities	[472]
	Antitumour activity	on lung metastasis	[473]
	Anti-AIDS virus activity		[474]
	Treatment of severe/cerebral malaria		[475]
Cellulose	Anticoagulant properties		[476]
	Influence on the blood pressure		[477]
	Treatment of periodontitis		[478]
	Anti-AIDS virus activity		[479]
Dextran	Anticoagulant properties		[480]
	Anti-AIDS virus activity		[479]
Xylan	Anticoagulant properties		[480]
	Antitumour activity	Inhibitory activities on lung metastasis	[473]
Schizophyllan	Anti-AIDS virus activity	Suppressing proliferation of AIDS virus	[481]
Chitin	Inhibition of cell proliferation		[482]
Chitosan			[483]
Palmitoyldextran phosphates	Antitumour activity	82% growth regression against sarcoma 183 ascites-tumour	[484]
Phenylacetate carboxymethyl benzylamide dextran	Aponecrotic, antiangiogenic and antiproliferative effects on breast cancer growth	The dextran ester acts like the pure components but in more potent manner	[485]
Xylan[a]	Anticoagulant properties		[350]
Dextran[a]	Immunostimulating properties		[347]
Chitosan[a]	Bone repair		[486]
	Anti-inflammatory effects		[487]

[a] Phosphate

azidothymidine through enzymatic hydrolysis, a long hydrophobic alkylene is inserted between the drug and the backbone (Fig. 10.4), which leads to an increase in anti-HIV activity and a decrease in anticoagulant activity [491].

10.3 Carrier Materials

Drug targeting with novel medical devices is one of the major developments in the field of therapeutics. In addition to controlled release mechanisms accom-

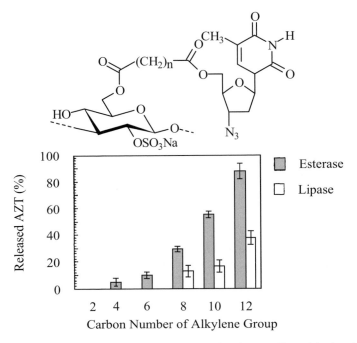

Fig. 10.4. Percentage of released azidothymidine from curdlan sulphuric acid half ester by esterase and lipase catalysis at 37 °C for 24 h (reproduced with permission from [491], copyright American Chemical Society)

plished with defined coatings (Fig. 10.1), it is the development of nanoparticle systems usable for parental injection, the use of hydrogels, and the preparation of hydrolytically instable prodrugs that are of recent importance. Because of the fact that polysaccharide esters can provide biocompatible and biodegradable materials, they are being studied increasingly.

10.3.1 Prodrugs on the Basis of Polysaccharides

Prodrugs are soluble polysaccharide derivatives containing covalently bound bioactive agents that may be liberated in an organism by defined cleavage of the covalent bond. Polysaccharide esters are well suited for this because of the ease of deesterification by simple hydrolysis or enzymatic attack. A broad variety of drugs can be bound, and the polysaccharide of choice is almost exclusively dextran to yield water-soluble prodrugs.

Covalent coupling of methotrexate to dextran enhances the penetration of cytotoxic substances into a tissue-like matrix [492]. Experiments towards the ability of these prodrugs to kill cells and to penetrate through tissue such as human brain tumour (H80) cells reveal that slowly eliminated agents can penetrate further through tissues and that the dose-response curve is shifted to lower dosage.

Studies on the dextran ester of the antiasthmatic drug cromoglycic acid indicate that the cromoglycate is released with a half-life of 10 h if the acylation is carried out with the chloride of the drug, yielding a loading of 2.5% (w/w). If the ester is prepared via the imidazolide, the product contains between 0.8 and 40% (w/w) of the cromoglycate, depending on the reaction conditions. An ester with 0.8% (w/w) releases the cromoglycate with a half-life of 39 min, while another batch containing 40% (w/w) cromoglycate has a release half-life of 290 min in buffer of pH 7.4 at 37 °C [230]. The hydrolysis of dextran metronidazole succinate over the pH range 7.4–9.2 at 37 °C can be determined with HPLC, and shows slower release compared to the cromoglycate. Interestingly, an intramolecularly catalysed hydrolysis by the neighbouring dextran hydroxy groups is observed [493]. For the dextran metronidazole esters, in which succinic and glutaric acids are incorporated as spacers, the decomposition proceeds through parallel formation of metronidazole and the monoester derivative, as can be demonstrated by reversed-phase HPLC and SEC. Almost identical stability of the individual esters is obtained after incubation in 0.05 M phosphate buffer pH 7.40 and in 80% human plasma, revealing that the hydrolysis in plasma proceeds without enzymic catalysis. The half-lives of the polymeric derivatives derived from maleic, succinic and glutaric acids are 1.5, 32.1 and 50.6 h respectively [494]. A metronidazole-containing prodrug is also achieved by acylation of inulin with metronidazole monosuccinate in the presence of DCC [222].

Dextran nalidixic acid ester with different DS values has been synthesised as a colon-specific prodrug. The water-soluble derivatives show no drug release for 6 h under conditions similar to those of the stomach, indicating that the prodrug is chemically stable during the transit through the gastrointestinal tract [495]. Comparable behaviour is observed for the ester prepared as a colon-specific prodrug of 5-aminosalicylic acid, which is active against inflammatory bowel diseases, and for dextran-5-(4-ethoxycarbonylphenylazo)salicylic acid ester [496, 497].

The preparation of prodrugs is a valuable approach for the transportation of lipophilic agents in a biological environment. This is well illustrated by the lipophilic agent naproxen, which can be bound to dextran via acylation. The water solubility of naproxen bound to dextran is sometimes higher than for the acid form (up to 500 times). At 60 °C, the hydrolysis of the ester in aqueous buffer solution is acid base catalysed. An almost identical degradation rate is obtained for the ester in 80% human plasma, excluding catalysis of hydrolysis by plasma enzymes [498, 499]. Dextran esters of ketoprofen, diclofenac, ibuprofen and fenoprofen have been studied, showing that the dextran ester prodrug approach provides selective colon delivery systems of drugs possessing a carboxylic acid functional group [500, 501].

Water-soluble sulphonylurea pullulan esters are used to study the insulinotropic activity and cell viability, by means of rat pancreatic islets co-entrapped with the polysaccharide derivative in conventional alginate-poly(L-lysine) microcapsules. A long-term (1 month) culture experiment has revealed that the microcapsules of islets with sulfonylurea pullulan esters, with well-preserved morphology, present higher insulin secretion level and better ability in responding to glucose changes than those without the esters [502].

10.3.2 Nanoparticles and Hydrogels

The tendency of polysaccharides and their derivatives to form defined super-structures has been exploited in the preparation of polymeric nanostructures. Polymeric nanoparticles have gained considerable interest in the medical field, especially as carriers for drug targeting suitable for parental injection, multifunctional biomedical devices, and as contrast enhancers [218, 503–505]. Synthesis strategies are known to gain partially hydrophobised polysaccharide derivatives capable of nanoparticle formation. One path described in Sect. 5.2 is the acylation reaction of remaining OH functions in partially substituted pullulan acetate with carboxymethylated poly(ethyleneglycol) using DCC. A particle size of about 193 nm and a unimodal distribution have been demonstrated by means of photon correlation spectroscopy (PCS). Clonazepam, as a model drug, is easily incorporated by the polymeric nanoparticles and shows drug release behaviour controlled mainly by diffusion from the core portion [218]. In addition, enzyme-catalysed transesterification is used for the selective esterification of starch nanoparticles with vinyl stearate, applying *Candida antarctica* Lipase B as catalyst. After removal of the surfactant from the modified starch nanoparticles, they can be dispersed in DMSO or water and retain their nanodimensions [246].

Adjustment of the hydrophilic–hydrophobic balance necessary for the formation of polymeric nanoparticles via micelle formation in a dialysis process is achieved by a defined two-step esterification of dextran with biocompatible propionate and pyroglutamate moieties, leading to highly functionalised derivatives [506]. The major fraction consists of particles of 370 nm in diameter (Fig. 10.5a). The SEM image in Fig. 10.5b demonstrates that nanospheres in the suspensions do not undergo any morphological changes within 3 weeks. Nanoparticles based on chitosan can be accomplished by chemoselective conversion of chitosan with deoxycholic acid in methanol, as described in Chap. 9, yielding particles with diameters in the range 161–180 nm [507].

In addition to nanoparticles, hydrogels are advanced polysaccharide-based materials for drug delivery systems and protective encapsulants, e.g. of viruses used in gene therapy [508]. A promising polysaccharide ester in this regard is dextran maleic acid monoester. The hydrogel precursor is soluble in various, common organic solvents. The hydrogels are made by the irradiation of dextran maleate with long-wave UV light (365 nm), and show a high swelling (swelling ratio from 67 to 227%) depending on DS and the pH of the medium, i.e. highest swelling ratio in neutral pH, followed by acidic (pH 3) and alkaline conditions (pH 10). The swelling ratio increases with an increase of the DS [509]. The surface and interior structure of a dextran methacrylate hydrogel prepared in a comparable manner has been investigated by means of SEM after application of special cryofixation and cryofracturing techniques. A unique, three-dimensional porous structure is observed in the swollen hydrogel (Fig. 10.6), which is not evident in the unswollen

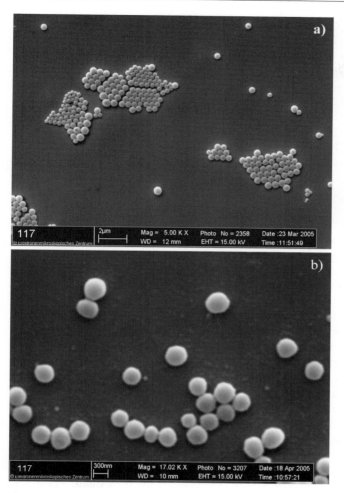

Fig. 10.5. SEM images of dextran propionate pyroglutamate nanoparticles on a graphite-covered mica surface taken **a** directly after the dialysis and **b** after 3 weeks storage in water (reproduced with permission from [506], copyright American Chemical Society)

hydrogel. Different pore sizes and morphologies between the surface and the interior of swollen hydrogels are visible [156].

Transparent hydrogels useful for adhesion inhibitors, tissue adhesives, wound dressings, hemostatics, and embolisation materials are obtained from dextran methacrylates via polymerisation with N-isopropylacrylamide in DMSO in the presence of azobisisobutyronitrile [510].

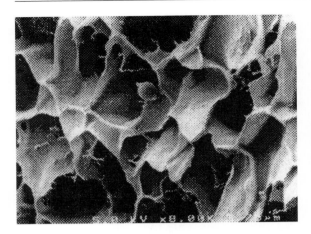

Fig. 10.6. Three-dimensional porous structure of a dextran-methacrylate hydrogel observed by means of SEM (Kim S-H, Chu C-C, Synthesis and characterization of dextran-methacrylate hydrogels and structural study by SEM, J Biomed Mater Res 49, pp 517, copyright (2000) American Association for the Study of Liver Diseases, reprinted with permission of Wiley VCH)

10.3.3 Plasma Substitute

A promising application is plasma substitutes based on polysaccharide esters, in particular, starch acetate. Colloidal plasma substitutes are aqueous solutions of colloids and electrolytes isotonic with blood, which are used to replace lost blood. They stabilise the circulation, dilute the blood (haemodilution) and improve the microcirculation. There is a growing market for man-made colloidal plasma substitutes because they exclude the risk of disease transmission, ensure sterility, have a long shelf life, and can be cheaply prepared in large amounts [511]. Plasma substitutes need to have the same colloid-osmotic pressure as the original blood and, therefore, they are usually polymeric substances. The nowadays broadly applied hydroxyethyl starch has a number of drawbacks. The degradation is slow and incomplete because etherification hinders the attack of the α-amylase. Prolonged application of hydroxyethyl starch may lead to accumulation in the human body.

Starch acetate is quickly metabolised because the ester moiety is attacked by an esterase, and the liberated starch can be degraded by the α-amylase. The starch starting material needs to have a molecular mass in the range of 10 000–500 000 g/mol. This can be achieved by defined enzymic degradation [130,512,513]. The starch acetates applied have DS values in the range 0.3–0.8, and acetylation of all the reactive sites has been demonstrated by NMR spectroscopy. The substitution is in the order position 2 = 3 > 6 [514]. Investigation on the behaviour of starch acetate in blood plasma of male volunteers exhibits a metabolic path different to that of hydroxyethyl starch, suggesting a more rapid and complete degradation of the starch acetate in the body.

The higher tendency of the ester bound towards hydrolysis, which is desirable in blood, raises the problem of a shorter shelf life. A selective acetylation at position 2 can overcome this problem because it leads to a more stable derivative, as shown in Fig. 10.7.

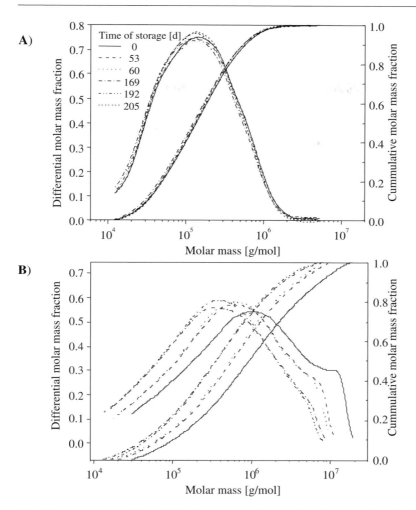

Fig. 10.7. Comparison of the stability (measured as decrease in molar masses) of selectively functionalised starch acetate (**A**) with the stability of non-selectively acetylated starch (**B**) over the course of 205 days

11 Outlook

Polysaccharides, as naturally occurring polymers, are by far the most abundant, renewable resource in the world, possessing magnificent structural diversity and functional versatility. These biopolymers are amongst the key substances that make up the fundamental components of life. Some of the polysaccharides – in particular, cellulose and starch and semisynthetic derivatives thereof – are actively used in commercial products today, while the majority of polysaccharides and alternative esters are still underutilised. Basic and applied research has already revealed much knowledge and, from the state of the art of polysaccharide esterification, further innovations and an increased use of the biopolymers may be expected.

Within the framework of this book, the structure analysis and esterification of cellulose, scleroglucan, schizophyllan, dextran, pullulan, starch (amylose, amylopectin), xylan, guar, curdlan, inulin, chitin, chitosan and alginate are described to illustrate typical features and procedures in the field of polysaccharide research and development. Readers will appreciate that the most important polysaccharide, as basis for chemical modification reactions and in terms of application, is cellulose, as cellulose is the most abundant polysaccharide and it occurs in a rather pure form. Moreover, cellulose and cellulose derivatives are among the oldest polymeric materials isolated and applied. Nevertheless, this picture may change because much effort is being invested today to provide a broader variety of polysaccharides in sufficient quantity and quality, which will be achieved by new biotechnological approaches and efficient purification techniques. The versatile structures of polysaccharides available will fertilise the utilisation of this important group of biomacromolecules, even surpassing the most important strategy of employing polysaccharides and their derivatives for green chemistry paths. It is not only the opportunity to include these biopolymers in biological cycles (cradle-to-grave approach) but also their amazing tendency to form architectures via supramolecular structures and to act in biological systems that will be undoubtedly lead to new applications.

To take full advantage of polysaccharides, tailored chemical modification is a powerful tool. Esterification of polysaccharides will surely continue to play an important role in the development of new products based on this important renewable resource. Structure and, hence, property design can be accomplished by choosing both the "right" polysaccharide and acid for the esterification. The routes for esterification described in this book involve covalent binding of virtually any acid known, at least in laboratory-scale chemistry. A number of state-of-the-art reagents of modern organic chemistry are already employed but the consequential search for esterification paths appropriate for polysaccharides, i.e. without any side reactions at the polymer backbone, is indispensable to provide highly pure and biocompatible materials, which may even enter the commercial scale. The usefulness of enzymic acylation reactions should be carefully investigated. Efficient esterification steps are the prerequisite for multiple functionalisation resulting in tailor-made products. Both subsequent reactions and one-pot conversion with different acids are of scientific and commercial interest, yielding derivatives with various patterns of functionalisation. An associated challenge is the chemoselective and regioselective esterification. New protective groups designed for polysaccharides will be a key development for progress in this field. For synthesis of new polysaccharide esters, one should always consider the utilisation of naturally occurring acids, not only to accommodate the green chemistry approach but also to follow the concept of biomimetics.

A fast-growing area is the structure elucidation of the broad variety of known and new polysaccharide derivatives. Among the most promising strategies for the analysis of the molecular structure of polysaccharide esters is NMR spectroscopy, in particular ^{1}H NMR spectroscopy after complete subsequent functionalisation, which combines efficiency, reliability and low costs with a maximum of structural information. Still necessary for making this path broadly applicable is a reasonable set of standard functionalisation procedures. Although the use of chromatographic techniques is a sophisticated business, it could provide information on the distribution of ester functions along the polymer backbone if appropriate subsequent modification and mathematic simulations are established. Solution of this scarcely treated problem is necessary for the establishment of general structure property relations. ^{13}C NMR and multidimensional NMR spectroscopy seem to be the methods of choice to gain information at the level of superstructure. Moreover, if the development of scanning probe microscopy continues with the breathtaking tempo of the last decade, then AFM or a comparable microscopic method could be able to simply visualise molecular and supramolecular structures of polysaccharides and polysaccharide esters.

The evaluation of recent literature in the field of polysaccharide esters revealed a change in trends for application. "Traditionally", the derivatives are employed according to their bulk properties, i.e. solubility, thermal properties, and film forming. New applications of highly engineered polysaccharide esters will focus much more on the assembly of supramolecular architectures and defined interaction, also via recognition processes. The naturally given structural features of the polysaccharides, such as the multiple chirality of the polymer backbone, will therefore be exploited to a much larger extent. Bioactivity, scarcely studied up to now but with an enormous potential for polysaccharide application, as well as the biodegradability of polysaccharide esters controlled by a designed functionalisation will also be an important research field. Biomimetic approaches can substantially broaden the use of these biopolymer derivatives because polysaccharide esters are predestined to reproducibly mimic natural structures.

12 Experimental Protocols

Cellulose triacetate, Dormagen method (adapted from [515, 516])

6 g (37 mmol) cellulose is mixed with 2.1 g (35 mmol) glacial acetic acid and kept for 18 h at RT: 16.5 g (0.162 mol) acetic anhydride, 22.0 ml dichloromethane and 0.03 ml concentrated H_2SO_4 are added and the temperature is kept below 25 °C. The fibre pulp is slowly heated (15 °C per h) under stirring to 40 °C and kept at this temperature until total dissolution of the fibres occurs. 2 g potassium acetate is dissolved in 50% aqueous acetic acid and added to the reaction mixture to decompose the excess H_2SO_4. 80.0 ml water is added drop-wise in order to convert the acetic anhydride to acetic acid. The methylene chloride is evaporated under vacuum, followed by pouring the residual viscous mass into water. The cellulose acetate is intensively washed with water and finally dried. Solubility, see Table 4.1.

Cellulose-2,5-acetate, secondary acetate (adapted from [516])

6.0 g (37 mmol) cellulose is ground with 2.1 g (35 mmol) glacial acetic acid and kept for 18 h at RT. 16.5 g (0.162 mol) acetic anhydride, 22.0 ml methylene chloride and 0.03 ml concentrated H_2SO_4 are added while the temperature is kept below 25 °C. The fibre pulp is slowly heated (15 °C per h) under stirring to 40 °C, and the temperature is maintained until total dissolution of the fibres. A solution of 0.03 ml concentrated sulphuric acid in 3.0 ml water is added and stirring at 60 °C is continued until the cellulose ester is soluble in acetone. This can be tested by precipitation of a small sample in methanol, washing with methanol and then testing the solubility. When the cellulose ester is acetone soluble, a mixture of 0.6 g potassium acetate dissolved in 50% acetic acid is added. After evaporating the dichloromethane under vacuum, the cellulose ester is isolated by precipitation in water, washing and drying. DS 2.38, solubility, see Table 4.1.

Cellulose triacetate, polymeranalogues reaction (adapted from [88])

10 g (62 mmol) cotton linters is placed in a 250 ml conical flask, followed by a mixture of 80 ml (1.4 mol) glacial acetic acid, 120 ml toluene and 2.0 ml 71–73% $HClO_4$. The mixture is shaken vigorously for a few minutes, and the excess liquid is decanted into 50 ml (0.529 mol) acetic anhydride. This mixture is swirled and immediately poured back into the flask containing the linters. The purpose of this

procedure is to minimise the possibility of a high initial concentration of acetic anhydride contacting the fibres closest to the flask, and hence generating material that might have higher than average degrees of acetylation. The closed flask is shaken for 8 h at 30 °C. The acetylated linters are removed, and washed three times with ethanol and then several times with water in order to remove residual traces of acid. Washing is continued until the solution is neutral. The cellulose acetate is washed with ethanol and subsequently dried under vacuum overnight at 60 °C. DS 2.93, ^{13}C NMR (DMSO-d_6): δ (ppm) = 171.3, 170.4, 170.1 (C=O), 100.4, 77.0, 73.6, 73.2, 72.9, 63.4 (polymer backbone), 21.2, 20.9 (CH$_3$).

Cellulose valerate, heterogeneous reaction in Py and activation of the carboxylic acid with TFAA as impeller (adapted from [95])

100 ml (19.4 mol/mol AGU) TFAA and 96 ml (23.9 mol/mol AGU) valeric acid are mixed and stirred at 50 °C for 20 min. This solution is added to 6.0 g (37 mmol) dried cellulose powder and heated at 50 °C for 5 h. The reaction mixture is poured into methanol and the polymer is filtered off, washed repeatedly with methanol and dried. DS 2.79.

Hemicellulose acetate, synthesis in DMF and activation with NBS (adapted from [98])

0.66 g dry hemicellulose powder dispersed in 10 ml distilled water is heated to 80 °C under stirring until complete dissolution (~ 10 min). 5 ml DMF is added and the mixture stirred for 5 min. The water is removed from the swollen gel by repeated azeotropic distillation under reduced pressure at 50 °C for 0.5 h. In this case, about 12 ml solvent is recovered. 30 ml acetic anhydride and 0.3 g (1.3 mmol) NBS are added and the homogeneous reaction mixture is heated at 65 °C for 5 h. After cooling to RT, the mixture is slowly poured into 120 ml ethanol (95%), with stirring. The product is filtered off, washed thoroughly with ethanol and acetone, and dried initially in air for 12 h and subsequently for 12 h at 55 °C. DS 1.15. FTIR (KBr): 1752 ν (C=O), 1347 ν (C–CH$_3$), 1247 ν (C–O) cm^{-1}.

Dextran palmitate, heterogeneous synthesis in Py/toluene with palmitoyl chloride (adapted from [92])

20 g (123 mmol) dextran, 100 g (363 mmol) palmitoyl chloride, 75 g Py and 75 g toluene are heated under reflux with vigorous agitation at 105–110 °C for 1.5 h. The mixture is cooled rapidly to RT and washed with 250 ml water. About 100 ml of chloroform is added to the residue, followed by shaking. The resulting solution is poured into 1 l methanol to precipitate the ester. The ester is filtered off, redissolved in a mixture of 75 ml toluene and 100 ml chloroform, reprecipitated in methanol, collected and dried. DS 2.9. The product is soluble in chloroform, carbon tetrachloride, benzene, toluene and xylenes.

Pullulan nonaacetate, heterogeneous synthesis in Py (adapted from [102])

1.0 g (6.2 mmol AGU, 1.8 mmol RU) pullulan is suspended in 20 ml Py, and 0.25 g (2.0 mmol) DMAP is added. The mixture is stirred for 2 h at 100 °C. 10 ml (0.106 mol) acetic anhydride is added and stirring is continued for 1 h. The product is precipitated in water, filtered off, and reprecipitated from acetone solution into water. DS 3.0. ^{1}H NMR (CDCl$_3$): δ (ppm) = 3.57 – 5.49 (H$_{Maltotriose}$), 1.96 – 2.14 (CH$_3$). ^{13}C NMR (CDCl$_3$): δ (ppm) = 62.8 – 96.0 (C$_{Maltotriose}$), 169.0 – 170.7 (C=O$_{Ester}$).

Chitin acetate, heterogeneous synthesis in Py (adapted from [103])

0.207 g (1.1 mmol) β-Chitin (DDA 0.16) is mixed in 20.0 ml (0.2473 mol) Py, and 10.0 ml (0.106 mol) acetic anhydride and 0.20 g DMAP are added. The mixture is stirred for 48 h at 50 °C under nitrogen. The resulting light reddish brown mixture is poured into ice water. The light tan fibrous precipitate is filtered off, washed with water and acetone, and dried. DS 3.0.

Starch octanoate, heterogeneous synthesis in Py with octanoyl chloride (adapted from [517])

To 2.5 g (15 mmol) dried starch are added 15 ml Py and 45 g (0.28 mol) octanoyl chloride, and the reaction is heated for 6 h at 115 °C. The mixture is cooled and poured into 200 ml absolute ethanol, with vigorous stirring. The product is filtered off and washed twice with 200 ml ethanol. Excess ethanol is removed by an air stream and the starch ester is dried at 50 °C overnight. DS 2.7. FTIR (KBr): 3380 ν (OH), 2927 and 2856 ν (CH), 1746 ν (C=O$_{Ester}$) cm^{-1}. ^{1}H NMR (CDCl$_3$): δ (ppm) = 0.81 (CH$_3$), 1.20 (CH$_2$), 1.43 (COCH$_2$CH$_2$), 2.26 (OCOCH$_2$), 3.50 – 5.40 (H$_{AGU}$).

Cellulose, dissolution in DMAc/LiCl (adapted from [169])

1.0 g (6.2 mmol) dried cellulose and 40.0 ml DMAc are heated for 2 h at 130 °C under stirring. After cooling to 100 °C, 3.0 g anhydrous LiCl is added. The cellulose dissolves completely by cooling to RT under stirring.

Cellulose, dissolution in DMSO/paraformaldehyde (adapted from [193])

0.10 g (0.62 mmol) cellulose and 0.50 g paraformaldehyde are dispersed in 10.0 ml DMSO and heated, with rapid stirring, to 130 °C for 6 – 8 min. Evolution of formaldehyde occurs and, shortly after the onset of vigorous bubbling, a clear solution is obtained. If the cellulose does not dissolve, the amount of paraformaldehyde may be increased to achieve complete solution. The water content of the paraformaldehyde used should not exceed 5% and the water content of the entire DMSO/paraformaldehyde/cellulose system should be less than 1%.

Cellulose acetate nonanoates, heterogeneous synthesis in DMAc and titanium(IV)isopropoxide as catalyst (adapted from [518])

5.0 g (31 mmol) cellulose and 35 g DMAc are heated to 100 °C under nitrogen for 1 h. 6.3 g (2 mol/mol AGU) acetic anhydride, 19.4 g (2 mol/mol AGU) nonanoic anhydride and 0.15 g (17.1 mmol/mol AGU) titanium(IV)isopropoxide are added to the activated cellulose, and the mixture is heated to 150 °C for 8 h. After cooling to 20 °C, the clear solution is poured into methanol. The product is filtered off and slurried in methanol. This process is repeated until the filtrate becomes clear. The slurry and filtration process is repeated twice with water. The product is dried in a vacuum oven under nitrogen at 60 °C. DS_{Acetyl} 2.03, $DS_{Nonanoyl}$ 0.70. The product is soluble in acetone, acetic acid, THF, $CHCl_3$, DMSO and NMP.

Starch heptanoate, acylation in water with heptanoyl chloride (adapted from [136])

6.75 g (42 mmol) starch (Hylon VII) is added to 50 ml 2.5 M aqueous NaOH solution at RT with mechanical stirring under N_2, until the starch granules are fully gelatinised (about 30 min). 3.1 g (21 mmol) heptanoyl chloride is added drop-wise and the reaction mixture is stirred for 1 h. After neutralisation to pH 7 with acetic acid, the product is precipitated with 150 ml methanol. It is collected by filtration, washed with 150 ml aqueous methanol (70%) and filtered off, and the process is repeated twice. Methanol is removed by evaporation in air and the starch ester is dried at 50 °C overnight. DS 0.25.

Dextran acetate, homogeneous synthesis in formamide with acetic anhydride (adapted from [519])

To 1.0 g (6.2 mmol) dextran, 15 g formamide and 15 g Py are added with stirring followed by the slow addition of 12 g acetic anhydride. The solution is stirred for 18 h at 20–30 °C. The reaction mixture is poured slowly into 150 ml water. The dextran acetate is removed by centrifugation and washed with water. The product is slightly coloured. A nearly colourless polymer is obtained by finally washing with ethanol. DS 3.0. The product is soluble in tetrachloroethane.

Pullulan monosuccinate, homogeneous synthesis in DMSO with succinic anhydride (adapted from [144])

To a solution of 1.0 g (6.2 mmol) pullulan in 15 ml DMSO, 2.5 g (25 mmol) succinic anhydride in 10 ml DMSO is added. The solution is warmed to 40 °C and 0.28 g (2.6 mmol) DMAP is added, with stirring. The product is isolated by precipitation in a fivefold volume of an ethanol:ether mixture (1:1, v/v) and collected. The dried precipitate is dissolved in 10 ml water and purified by preparative gel filtration (300 ml, Sephadex G-2, eluent water, flow rate 2 ml/min, detection, 400 refractive index). The polymer fraction is collected and freeze dried. DS 3.0.

Cellulose acetate, homogeneous synthesis in [C₄mim]Cl (adapted from [153])

0.5 g (3.09 mmol) cellulose is mixed with 4.5 g molten [C₄mim]Cl and stirred for 12 h at 10 °C above the melting point of [C₄mim]Cl until complete dissolution. 1.09 ml (5 mol/mol AGU) acetyl chloride is added dropwise at 80 °C and stirred for 2 h. Isolation is carried out by the product being precipitated in 200 ml methanol, washed with methanol and dried under vacuum at 60 °C. Yield: 0.75 g (85.9%). DS 3.0. IR (KBr): 2890 ν (CH), 1750 ν (C=O$_{Ester}$) cm^{-1}. ^{13}C NMR (DMSO-d_6): δ = 168.9–170.2 (C=O), 62.1–99.2 (modified AGU).

Dextran maleate, homogeneous synthesis in DMF/LiCl with maleic anhydride (adapted from [509])

5.0 g (30.8 mmol) dextran is dissolved in DMF/LiCl (50 ml/4.5 g) at 90 °C under nitrogen, cooled to 60 °C and 0.4 ml (3 mol % to maleic anhydride) TEA is added. The solution is stirred for 15 min at 60 °C and then 9.0 g (3 mol/mol AGU) maleic anhydride is added slowly. The mixture is stirred for 16 h at 60 °C under nitrogen. The polymer is precipitated with cold isopropanol, filtered off, washed several times with isopropanol, and dried under vacuum at RT. DS 0.84. FTIR (KBr): 3500–2500 ν (OH), 3052 ν (C=C–H), 1728 ν (C=O), 1660–1640 ν (C=C), 824–822 δ (C=C–H) cm^{-1}. ^{1}H NMR (DMSO-d_6): δ (ppm) = 7.5 and 6.2 (C=C–H). ^{13}C NMR (DMSO-d_6): δ (ppm) = 137 and 131 (C=C).

Dextran α-naphthylacetate, homogeneous synthesis in DMF/LiCl and in situ activation with TosCl (adapted from [202])

2.0 g (12.3 mmol) dextran is dissolved in 100 ml DMF containing 2.0 g LiCl at 90 °C, cooled to 50 °C, and 5.85 g (74 mmol) Py, 6.89 g (37 mmol) α-naphthylacetic acid and 37 mmol TosCl are added, with stirring. After 22 h, the modified polymer is precipitated with isopropanol/diethylether. The samples are purified by reprecipitation from THF solution in isopropanol/diethylether and then dried under vacuum over P₂O₅. DS 0.56. FTIR: 1720 ν (C=O), 1590 and 1510 ν (C–C$_{Aromat}$) cm^{-1}. ^{1}H NMR (DMSO-d_6): δ (ppm) = 7.8–7.2 (H$_{Aromat}$), 4.1 (CH₂), 3.6–5.4 (dextran backbone). ^{13}C NMR (DMSO-d_6): δ (ppm) = 170.3 (C=O), 133–123.4 (C$_{Aromat}$), 98.1–66.6 (dextran backbone), 37.5 (CH₂).

Cellulose adamantate, homogeneous synthesis in DMAc/LiCl with adamantoyl chloride (adapted from [169])

3.7 g (18.6 mmol) adamantoyl chloride and 1.8 ml (22.3 mmol) Py are added to 40 ml of a solution containing 2.5% (w/w, 6.2 mmol) cellulose and 7.5% LiCl in DMAc and stirred for 24 h at 80 °C. The homogeneous reaction mixture is poured into 250 ml ethanol. After filtration, the polymer is washed with ethanol and dried under vacuum at RT. DS 1.92. FTIR (KBr): 3457 ν (OH), 2909, 2854 ν (CH), 1720 ν (C=O$_{Ester}$) cm^{-1}. ^{13}C NMR (CDCl$_3$): δ (ppm) = 176.5 (C=O), 103.0 (C-1), 100.9 (C-1′), 81.3 (C-2,3$_s$, C-4), 77.0 (C-3, C-5), 73.6 (C-2), 61.2 (C-6$_s$), 40.9 (α-C), 39.0 (β-CH$_2$), 36.4 (δ-CH$_2$), 27.9 (γ-CH).

Cellulose acetate, homogeneous synthesis in DMSO/TBAF with acetic anhydride (adapted from [129])

To 1.0 g (6.2 mmol) cellulose dissolved in 60 ml DMSO containing 11% (w/v) TBAF is added 1.45 ml (15.3 mmol) acetic anhydride. The mixture is heated for 3 h at 60 °C and the product is precipitated with 250 ml methanol, filtered off, washed with 50 ml methanol, and dried under vacuum at 50 °C. DS 1.20. FTIR (KBr): 1752 ν (C=O) cm^{-1}. ^{13}C NMR (DMSO-d_6): δ (ppm) = 169.1–169.9 (C=O), 60.3–102.5 (cellulose backbone).

Cellulose triacetate, homogeneous synthesis in 1-ethyl-pyridinium chloride/Py (adapted from [123])

100 g 1-ethyl-pyridinium chloride is mixed with 50.0 ml Py and melted at 85 °C. 2.0 g (12.3 mmol) cellulose is added, with stirring. After 1 h, a clear solution is obtained. A solution of 38 ml (0.402 mol) acetic anhydride in 55 ml (0.682 mol) Py is added and heated for 30 min with stirring at 80 °C. After 20–30 min, the cellulose derivative precipitates as flakes. After cooling to 40 °C, dichloromethane is added until a clear solution is formed. The product is isolated by precipitation in methanol. The polymer is collected, washed with methanol, and dried. It can be further purified by reprecipitation from dichloromethane in methanol. DS 3.0.

Cellulose furoate, homogeneous synthesis in DMAc/LiCl (adapted from [170])

2 g (12.3 mmol) cellulose is activated with 200 ml distilled water for 48 h, followed by filtration. The polymer is stirred in 100 ml DMAc for 24 h and filtered off. This procedure is repeated three times. 2 g activated cellulose is dissolved in 100 ml solution of 9% LiCl in DMAc. 2.5 ml (2.5 mol/mol AGU) Py in 25.0 ml DMAc is added slowly to the cellulose solution, followed by 6.0 ml (5 mol/mol AGU) 2-furoyl chloride in 25.0 ml DMAc drop-wise. After stirring for 5 h at RT, the product is precipitated as a white powder by pouring the solution into hot distilled water. It is separated, washed several times with water, Soxhlet extracted using methanol for 24 h, and freeze dried. DS 2.5.

Cellulose trifluoroacetate (adapted from [187])

1.0 g (6.2 mmol) cellulose is swollen in 20 ml (269 mmol) TFA at RT for 20 min. 10 ml (72 mmol) TFAA is added to the slurry. The cellulose dissolves completely after stirring for 4 h at RT. The solution is stirred for an additional 2 h and poured into 200 ml diethyl ether. The polymer is filtered off, washed with 100 ml diethyl ether, and dried under vacuum for 20 h at 25 °C. In order to remove traces of both diethyl ether and trifluoroacetic acid, the sample is treated under vacuum for 40 min at 150 °C. DS 1.5. FTIR (KBr): 1790 ν (C=O) cm^{-1}. The polymer is soluble in DMF, DMSO and Py.

Cellulose formate (adapted from [187])

1.0 g (6.2 mmol) cellulose is swollen in 30 ml formic acid for 15 min at RT, and 2.7 ml POCl$_3$ is added to the slurry at 5 °C. After stirring for 5 h at RT, the cellulose dissolves completely and the polymer is precipitated with 100 ml diethyl ether, filtered off, and washed three times with 350 ml acetone. After drying at RT, the polymer is washed again with 200 ml acetone and dried under vacuum at 25 °C for 24 h. DS 2.2. FTIR (KBr): 1728 ν (C=O) cm^{-1}.

Cellulose laurate, homogeneous synthesis in DMAc/LiCl and in situ activation with TosCl (adapted from [127])

2.38 g (12.5 mmol) TosCl is added to 40 ml of a solution containing 2.5% (w/w, 6.2 mmol) cellulose and 7.5% LiCl in DMAc, followed by 2.47 g (12.5 mmol) lauric acid, under stirring. The reaction mixture is stirred for 24 h at 80 °C under N$_2$. The polymer is precipitated in 800 ml buffer solution (7.14 g K$_2$HPO$_4$ and 3.54 g KH$_2$PO$_4$ per 1 l H$_2$O) and then collected by filtration. After washing three times with 800 ml water and Soxhlet extraction with ethanol (24 h), it is dried under vacuum at 50 °C. DS 1.55. FTIR (KBr): 3486 ν (OH), 2925, 2855 ν (CH), 1238 ν (COC$_{Ester}$), 1753 ν (C=O$_{Ester}$) cm^{-1}. ^{13}C NMR (CDCl$_3$): δ (ppm) = 173.8 (C=O), 104.0 (C-1), 102.6 (C-1'), 72.3 (C-2), 73.3 (C-3), 82.0 (C-4), 75.1 (C-5), 20.6–34.0 (C$_{Methylene}$), 13.9 (C$_{Methyl}$).

Cellulose 3,6,9-trioxadecanoate, homogeneous synthesis in DMAc/LiCl and in situ activation with TosCl (adapted from [199])

To a solution of 5.0 g (30.8 mmol) cellulose and 10 g LiCl, 200 ml in DMAc is added a solution of 5.8 g (30.4 mmol) TosCl in 20.0 ml DMAc. After stirring for 30 min at RT, a mixture of 11.6 g (60.8 mol) TosCl and 16.5 g (85.8 mol) 3,6,9-trioxadecanoic acid in 40.0 ml DMAc are added. The homogeneous reaction mixture is stirred for 3 h at 65 °C. After cooling to RT, the solution is poured into 1 l isopropanol, the precipitate is filtered off, washed with isopropanol, and dried under vacuum at 70 °C. DS 0.62. FTIR (KBr): 3440 ν (OH), 2920, 2888 ν (CH), 1753 ν (C=O) cm^{-1}. ^{13}C NMR (D$_2$O): δ (ppm) = 102.9 (C-1), 100.4 (C-1'), 79.6 (C-3$_s$, 2$_s$), 75.9 (C-4), 74.5 (C-4'), 73.5–72.8 (C-2, 3, 5), 71.6 (C-8), 71.0–68.3 (C-10–14), 63.8 (C-6$_s$), 60.7 (C-6), 58.6 (C-16).

Cellulose anthracene-9-carboxylate, homogeneous synthesis in DMAc/LiCl and in situ activation with TosCl (adapted from [207])

2.0 g (12.3 mmol) cellulose is suspended in 50 ml DMAc and stirred for 2 h at 120 °C, with exclusion of moisture. After cooling to 100 °C, 3.0 g LiCl is added. The mixture is stirred at RT until formation of a clear solution, and 7.03 g (36.9 mmol) TosCl is dissolved and 8.20 g (36.9 mmol) anthracene-9-carboxylic acid is added. After stirring for 4 h at 50 °C, the product is precipitated in 400 ml ethanol. The separated polymer is carefully washed with ethanol and dried under vacuum with increasing temperature up to 40 °C. DS 0.52. FTIR (KBr): 3422 ν (OH), 2882 ν (CH), 1790, 1723 ν (C=O) cm^{-1}. ^{1}H NMR (DMSO-d_6): δ (ppm) = 7.6–8.2 (H$_{Aromatics}$).

Cellulose long-chain carboxylic acid esters, heterogeneous synthesis in Py and in situ activation with TosCl (adapted from [198])

To 450 g (4.41 mol) Py, 15 g (93 mmol) cellulose and 105 g (0.551 mol) TosCl are added with stirring under nitrogen. The organic acid is slowly added to give a molar ratio TosCl/organic acid of 1:1. The mixture is stirred for 2 h at 50 °C, after which the fibres are filtered off, washed with ethanol, and soxhlet extracted with methanol for 2 h. Soxhlet extraction is continued for 12 h with fresh methanol. Subsequently, the fibres are washed with ethanol, dried with compressed air, and stored in a desiccator for 20 h at RT to evaporate the residual ethanol. DS$_{Undecylenate}$ 1.11, DS$_{Undecanoate}$ 0.59, DS$_{Stearate}$ 0.19, DS$_{Oleate}$ 0.14.

Cellulose benzoate, homogeneous synthesis and in situ activation with CDI (adapted from [225])

50 ml DMAc containing 9% LiCl and solvent-swollen cellulose (1.6 g, 20 mmol hydroxyl groups) is stirred until complete dissolution of the polymer. 8.25 g (40 mmol) DCC is added, followed by 4.88 g (40 mmol) benzoic acid and 0.05 g (4 mmol) DMAP. The product is isolated by filtration of the dicyclohexyl urea from the reaction mixture and precipitation of the supernatant into a 50:50 mixture of methanol and deionised water. The polymer is washed with methanol. In addition, the filtrate can be dialysed against DMF after dilution with DMF. The resulting product is obtained after removal of the solvent and vacuum drying at RT. DS 0.33. FTIR (KBr): 3500–3100 ν (OH), 3000–2900 ν (C–H$_{Aromat}$), 1675–1600 ν (C=O$_{Ester}$), 1550 ν (C=C$_{Aromat}$) cm^{-1}. ^{1}H NMR (DMSO-d_6): δ (ppm) = 3.3–5.5 (CH$_2$ and CH of cellulose), 6.9–8.0 (H$_{Aromat}$).

Starch poly(N,N-dimethylglycinate), homogeneous synthesis in DMSO and in situ activation with N,N-diisopropylcarbodiimide (adapted from [227])

To a solution of 1.0 g (6.2 mmol) wheat starch and 0.29 g (2.8 mmol) N,N-dimethylglycine in 40.0 ml dry DMSO, 1.52 ml (9.8 mmol) N,N-diisopropylcarbodiimide and 0.15 g (1.2 mmol) DMAP are successively added. The mixture is stirred overnight at RT under nitrogen and quenched by the addition of 1 ml water. The polymer is precipitated in 400 ml ethanol, filtered off, washed with acetone, and dried under vacuum at RT. DS 0.3. ^{1}H NMR (D$_2$O): δ (ppm) = 2.3 (CH$_3$), 3.2–5.5 (AGU).

Cellulose stearate, homogeneous synthesis in DMSO/TBAF and in situ activation with CDI (adapted from [405])

To a solution of 3.0 g CDI (18.5 mmol) in 30 ml DMSO is added 5.24 g stearic acid (18.5 mmol). After stirring overnight, 1.0 g cellulose (6.2 mmol) dissolved in 66 ml DMSO/TBAF (10%, w/v) is added. The mixture is stirred for 24 h at 80 °C under N$_2$. The homogeneous reaction mixture is poured into 500 ml ethanol and the polymer filtered off. After washing with 250 ml ethanol three times, the polymer is dried at 50 °C under vacuum. DS 1.35. FTIR (KBr): 3456 ν (OH), 2926 ν (C–H), 1238 ν (C–O–C$_{Ester}$), 1743 ν (C=O$_{Ester}$) cm^{-1}. ^{13}C NMR (DMSO-d_6): δ (ppm) = 173.5, 172.26 (C=O), 106.0 (C-1), 103.6 (C-1'), 73.2–77.2 (C-2, 3, 5), 81.8 (C-4), 62.64 (C-6$_s$), 22.6–33.9 (CH$_2$ stearate), 14.0 (CH$_3$ stearate). ^{1}H NMR (CDCl$_3$): δ (ppm) = 0.90 (CH$_3$ stearate), 1.31–2.35 (CH$_2$ stearate), 3.49–5.03 (AGU).

Starch benzoate, heterogeneous synthesis in water and benzoic acid imidazolide (adapted from [138])

To a suspension of 1.0 g (6.2 mmol) starch in 1.25 ml water is added 0.07 g (0.4 mmol) N-benzoylimidazole. The pH value of 8.0 is maintained with 3.0% aqueous NaOH. The reaction mixture is stirred at RT until the pH value is stable for about 2 h. The starch ester is acidified with dilute H$_2$SO$_4$, filtered off and washed with water. DS 0.05.

Cellulose ester with carboxymethyl-β-cyclodextrin, homogeneous synthesis in DMAc/LiCl and in situ activation with CDI (adapted from [234])

Sodium carboxymethyl-β-cyclodextrin (CMCD, DS 0.66, 3.2 g, 15 mmol) is stirred with 20.0 ml methanolic HCl (20%, v/v) for 20 min at RT to partially convert the carboxylate into the free acid form. The sample is isolated, washed with methanol and dried under vacuum at RT. To 3.0 g of this treated CMCD in 30.0 ml DMAc, 1.0 g (6.2 mmol) cellulose dissolved in DMAc/LiCl and 1.0 g (6.2 mmol) CDI in 10.9 ml DMAc are added. After stirring for 24 h at 80 °C, the mixture is poured into 200 ml ethanol. After filtration and washing with aqueous ethanol, the product is dried under vacuum at RT. FTIR (KBr): 3305 ν (OH), 2888 ν (CH), 1724 ν (C=O$_{Ester}$), 1655 ν_{as}(COO$_{Carboxylate}$), 1426 ν_s(COO$_{Carboxylate}$) cm^{-1}.

Starch fatty acid esters, homogeneous synthesis in DMSO and fatty acid imidazolides (adapted from [232])

To a solution of 6.50 g (40 mmol) dry amylomaize starch (70% amylopectin) in 80 ml dry DMSO, at 80 °C is added a few drops of a methanolic CH$_3$OK solution.

After stirring for 20 min, the temperature is increased to 90 °C and a solution of 20.0 g (80 mmol) dodecanoyl imidazolide in 100 ml DMSO is added over 1 h and heating continued for 3 h. The reaction mixture is poured into 300 ml water, the polymer is filtered off and washed several times with water. The crude product is crushed in a Waring blender and treated with water at 70 °C to remove remaining residues of imidazole. After drying at 55 °C overnight, the starch ester is purified by extraction with boiling isohexane. $DS_{Dodecanoate}$ 1.90, $DS_{Octanoate}$ 1.55, $DS_{Hexadecanoate}$ 1.66. FTIR (KBr): 3000–2850 ν (CH), 1740 ν (C=O) cm^{-1}. ^{1}H NMR (CDCl$_3$): δ (ppm) = 3.0–6.0 (AGU), 0.8–2.0 (CH$_3$).

Cellulose stearate, homogeneous synthesis in DMAc/LiCl with stearic anhydride and oxalyl chloride (adapted from [239, 520])

The formation of the iminium chloride and the conversion with the carboxylic acid are carried out as "one-pot reaction". 2.35 g (18.5 mmol) oxalyl chloride is added very carefully to 20 ml DMF at −20 °C. After the gas formation has ceased, 5.24 g (18.4 mmol) stearic acid is added. The mixture is added to a solution of 1.0 g (6.2 mmol) cellulose in DMAc/LiCl and heated at 60 °C for 16 h, during which gelation occurs. The cellulose ester floats and can be isolated simply by filtration and washing with ethanol. DS 0.63. FTIR (KBr): 3620, 2920, 1140 (cellulose backbone), 1745–1760 ν (C=O) cm^{-1}.

Inulin acrylate, acylation via enzyme-catalysed transesterification with vinyl acrylate (adapted from [244])

Synthesis of inulin ester by 30 mg/ml Proleather FP-F is performed in 15 ml anhydrous DMF containing 0.017 M (6.7%, w/v) inulin and 0.48 mol vinyl acetate/mol RU. The reaction mixture is shaken at 250 rpm and 50 °C in a temperature-controlled New Brunswick Scientific C24 orbital shaker (Edison, NJ) for 72 h. The reactions are terminated by removal of the enzyme, which is insoluble in DMF by centrifugation at 4000 rpm for 10 min. The supernatants are precipitated with acetone. The precipitate is subsequently dissolved in Milli-Q water and dialysed using a regenerated cellulose dialysis tube (molecular mass cut-off 1000 g/mol) for 2 days at 4 °C against water, followed by lyophilisation for 48 h. DS 0.34.

Starch 2-O-acetate, synthesis via enzyme-catalysed transesterification (adapted from [245])

2.0 g (12.3 mmol) starch (Hylon VII, 70% amylose) is dissolved in 40 ml DMSO at 80 °C over 15 min. After cooling, 2.3 mol/mol AGU vinyl acetate and 40 mg (2%, w/w) NaCl are added. The mixture is heated at 40 °C for 70 h, with slow stirring. The catalyst is removed by centrifugation and the product is precipitated in 400 ml isopropanol, filtered off, washed intensively with isopropanol, and dried under vacuum at 50 °C. DS 1.00 at position 2.

Cellulose acetoacetate, homogeneous synthesis in DMAc/LiCl via ring opening of diketene (adapted from [248])

8.0 g (49.3 mmol) cellulose and 300 ml DMAc are heated to 150 °C for 26 min in a round-bottom flask equipped with a short-path distillation apparatus. After addition of 15 g LiCl, the mixture is heated to 170 °C for 8 min and 77 ml distillate is collected at 170 °C. The reaction mixture is cooled to RT and allowed to stir overnight until formation of a crystal-clear solution. The solution is heated to 110 °C and 3.73 g (44.4 mmol) diketene is added drop-wise. Stirring is continued for 30 min at 110 °C. After cooling to RT, the product is precipitated with methanol, filtered off, Soxhlet extracted with methanol, and dried (DS 0.8).

Cellulose-(4-N-methylamino)butyrate hydrochloride, homogeneous synthesis in NMP/LiCl (adapted from [253])

To 25 ml of a solution of 1.0% (w/w) cellulose and 9% LiCl in NMP is added 1.4 ml (17 mmol) Py. A solution of (3.4 g, 18 mmol) TosCl in 5 ml NMP is added drop-wise. After 12 h, the product is precipitated with THF, hydrolysed in water, reprecipitated, and dried. DS 0.7. FTIR: 2810 v (NH), 1746 v (C=O) cm^{-1}. ^{13}C NMR (D$_2$O): δ (ppm) = 177.2 (C=O), 50.8, 35.5 and 23.3 (CH$_2$).

Acylation of pullulan by ring opening of lactones (adapted from [254])

0.5 g (3.09 mmol AGU, 1.03 mmol RU) pullulan is placed into a previously silanised ampoule. The ampoule, containing a stir bar, is silanised with chlorotrimethylsilane and dried for 24 h under vacuum (10 Torr) at 50 °C. The ampoule is removed under an argon blanket and covered with rubber septa. 14 ml DMSO is added to the ampoule via a syringe under an inert atmosphere to obtain a 3.5% (w/v) solution of pullulan. The ampoule is placed in an oil bath at 55 °C for 35 min to dissolve the polymer. After homogenisation, 410 µl (0.08 mmol) stannous octanoate in DMSO is added, followed by 2.0 ml (18 mmol) distilled ε-caprolactone. The ampoule is sealed under vacuum and placed in an oil bath for 6 days at 60 °C. The reaction mixture is poured into 150 ml cold methanol, filtered off, and Soxhlet extracted with chloroform for 48 h. The modified polysaccharide is dialysed against distilled water using a cellulose acetate membrane (Spectrum Medical, molecular weight cut-off 3000 g/mol) for 18 h. It is dried for 24 h under vacuum (1 mm Hg) at 50 °C over P$_2$O$_5$ (DS 0.10).

Cellulose tosylate, homogeneous synthesis in DMAc/LiCl (adapted from [257])

20.2 g (118.7 mmol) air-dry cellulose (Avicel®) is slurried in 470 ml DMAc and stirred for 1 h at 160 °C. Approximately 40 ml DMAc is subsequently removed by distillation under nitrogen atmosphere. 40 g anhydrous LiCl is added at 100 °C and complete dissolution of the polymer is achieved after stirring at RT. 59.4 ml (427 mmol) TEA in 40.6 ml DMAc are added while cooling the cellulose solution to 8 °C. 40.7 g (213 mmol) TosCl in 60 ml DMAc are added within 30 min, followed by further stirring the homogeneous solution for 24 h. The polymer is precipitated with 5 l ice water, collected, and washed carefully with 15 l distilled water and then suspended in 1 l acetone and precipitated in 3 l distilled water. After filtration and washing with ethanol, the sample is dried under vacuum at 50 °C. DS$_{Tos}$ 1.36. The polymer is soluble in acetylacetone, DMSO, DMA, DMF and dioxane. FTIR (KBr): 3523 ν (OH), 3072 ν (C−H$_{Aromat}$), 2891 ν (C−H), 1598, 1500, 1453 ν (C−C$_{Aromat}$), 1364 ν_{as}(SO$_2$), 1177 ν_s(SO$_2$), 814 δ(C−H$_{Aromat}$) cm^{-1}. ^{13}C NMR (DMSO-d_6): δ (ppm) = 21.2 (CH$_3$), 59.0–101.4 (cellulose backbone), 127.8–145.1 (C$_{Aromat}$).

Chitin tosylate, heterogeneous synthesis in Py (adapted from [103])

To a suspension of 0.20 g (1.0 mmol) β-chitin (squid, DDA 0.16) in 20 ml methanol is added 10 ml acetic acid anhydride and the mixture is stirred for 48 h at 40 °C under nitrogen. The resulting swollen mixture is poured into ice water to isolate the N-acetylated chitin. It is filtered off, washed with water and acetone, and dried to give a white fibrous material in quantitative yield. The DDA is 0.0, as determined by both conductometric titration and elemental analysis. FTIR (KBr): 3434 ν (OH and NH), 1655 ν (amide I), 1588 ν (amide II), 1155–1030 (pyranose) cm^{-1}.

Elemental analysis: calculated for C$_8$H$_{13}$NO$_5$*1.1 H$_2$O, 43.09% C, 6.87% H, 6.28% N, 43.01% C, 6.60% H, 6.22% N.

To a suspension of 0.201 g β-chitin, obtained after selective N-acetylation (DDA 0.0), in 10 ml Py is added 1.9 g (10 mol/mol repeating unit) TosCl and 2.0 g DMAP. The mixture is stirred for 48 h at RT under nitrogen to give a light yellow mixture. It is poured into ice water and the collected precipitate is washed with water and ethanol and dried. DS 0.83. FTIR (KBr): 3418 ν (OH and NH), 1662 ν(amide I), 1587 ν (C−C$_{Aromat}$), 1543 ν (amide II), 1176 ν (SO$_2$), 1120–1032 (pyranose), 815 δ(C−H$_{Aromat}$) cm^{-1}.

Chitin tosylate, homogeneous synthesis in DMAc/LiCl (adapted from [276])

3.0 g (14.8 mmol) chitin (DDA 0.04) is placed in a 500 ml three-necked flask equipped with nitrogen inlet/outlet and a magnetic stirrer. 100 ml 5% (w/w) solution of LiCl in DMAc solution is added, followed by stirring the mixture for 20 min at 10 °C under nitrogen. To 10 ml of chitin solution (0.3 g actual chitin weight), 7.2 ml TEA (51.9 mmol) is added, followed by a solution of 9.86 g (51.4 mmol) TosCl in 10 ml DMAc. The mixture is stirred for 24 h at 10 °C. The polymer is precipitated with cold water, filtered off and dried under vacuum. DS$_{Tos}$ 1.07. Elemental analysis: 47.87% C, 5.43% H, 3.89% N, 9.48% S.

Cellulose 5-*N,N*-dimethylamino-1-naphthalenesulphonic acid ester, homogeneous synthesis in DMAc/LiCl (adapted from [263])

2.0 g (12.3 mmol) cellulose (AVICEL®) is slurried in 30 ml DMAc and heated at 130 °C for 1 h. The mixture is allowed to cool to 100 °C, and 3.0 g LiCl is added. Dissolution of the polymer proceeds under further stirring overnight without heating. A solution of 5.16 ml (37.0 mmol) TEA in 10 ml DMAc is added drop-wise while the mixture is cooled to 10 °C. After addition of 5.0 g (18.5 mol) 5-*N,N*-dimethylamino-1-naphthalenesulphonic acid chloride in 15 ml DMAc, the mixture is stirred for 24 h. The polymer is precipitated in 500 ml methanol, filtered off, and washed with water and methanol. It was further purified by reprecipitation from DMF in methanol and dried under vacuum at RT. DS 0.67. Elemental analysis: 6.75% S. The product is soluble in DMAc, DMF, DMSO and Py. FTIR (KBr): 3465 ν (OH), 3085 ν (C−H$_{Aromat}$), 2949, 2889 ν (CH$_3$), 2836 ν (CH$_2$), 2791 ν (N−CH$_3$), 1358 ν_{as}(SO$_2$), 1177 ν_s(SO$_2$) cm^{-1}. ^{13}C NMR (DMSO-d_6): δ (ppm) = 45.6, 46.8 (C17), 61.1 (C-6), 67.9 (C-6$_s$), 75.6 (C-5), 77.1 (C-4'), 79.0 (C-4), 83.0 (C-3$_s$), 75.6 (C-3), 80.4 (C-2$_s$), 72.6 (C-2), 98.5 (C-1'), 103.2 (C-1), 116.3 – 151.0 (C-7–16). ^{1}H NMR (DMSO-d_6): δ (ppm) = 2.88 – 2.89 (CH$_3$), 7.29 – 8.58 (H$_{Aromat}$).

Cellulose sulphuric acid half ester, synthesis via trimethylsilylcellulose (adapted from [521])

To a solution of 15.6 g (42 mmol) trimethylsilylcellulose (DS 2.46) in 360 ml dry THF is added a solution of 17.36 g (113 mmol) SO$_3$-DMF complex in 100 ml dry DMF. After stirring for 2.5 h at RT, the mixture is poured into a solution of 10.7 g (267 mmol) NaOH in 2 l methanol. The precipitate is filtered off, washed carefully with methanol, dissolved in 500 ml water, and reprecipitated into ethanol. After filtration and washing with ethanol, the sample is dried under vacuum at 50 °C. DS 1.13. FTIR (KBr): 1240 ν (SO$_2$), 806 ν(SO) cm^{-1}. ^{13}C NMR (D$_2$O): δ (ppm) = 102.8 (C-1), 100.9 (C-1'), 79.0 – 73.3 (C-2, 3, 4, 5), 66.9 (C-6$_s$).

Cellulose sulphuric acid half ester, synthesis via cellulose acetate (adapted from [186])

2.0 kg (7.40 mol, DS 2.5) cellulose acetate is dissolved in 20 l dry DMF at 60 – 80 °C. Remaining water is subsequently removed by distillation of 5 l DMF under vacuum. After cooling to 20 °C, 1.05 kg (9.01 mol) ClSO$_3$H in 4 l dry DMF is added drop-wise. The reaction proceeds 2 h with stirring. After addition of an equimolar amount of sodium acetate trihydrate dissolved in a water/DMF mixture, the polymer is precipitated in 80 l aqueous sodium acetate (5%, w/v) and stirred for 30 min. For deacetylation, the product is slurried for 12 h at 20 °C in a mixture containing 800 g NaOH, 1.5 l water and 20 l ethanol. Subsequently, it is separated, washed initially with 90% and then with 96% ethanol, and dried at 80 °C to a constant mass. DS 0.32. Elemental analysis: 5.2% S, 6.12% Na.

2,3-*O*-Functionalised cellulose sulphuric acid half ester, synthesis via 6-*O*-protected cellulose (adapted from [439])

Sulphation of 6-mono-*O*-(4-monomethoxy)trityl cellulose

To a solution of 6.0 g (14 mmol) 6-mono-*O*-(4-monomethoxy)trityl cellulose (DS 0.98) in 150 ml DMSO is added 8.9 g (55.9 mmol) SO_3-Py complex and the mixture is stirred for 4.5 h at RT. The polymer is precipitated in 500 ml methanol, neutralised with ethanolic NaOH solution, washed with ethanol, and dried under vacuum at 50 °C. DS 0.99.

Detritylation

To a suspension of 2.0 g 6-mono-*O*-(4-monomethoxy)trityl cellulose sulphuric acid half ester in 250 ml methanol is added 12.0 ml concentrated HCl. The mixture is stirred for 16 h. The product is filtered off, suspended in water, and neutralised with aqueous NaOH, yielding a solution. After precipitation with ethanol, the product is dried under vacuum at 50 °C. DS 0.99. FTIR (KBr): 815 ν (S=O) cm^{-1}. ^{13}C NMR (D_2O): δ (ppm) = 102.4 (C-1'), 100.3 (C-1), 82.3 (C-3$_s$), 80.2 (C-2$_s$), 79.2–78.1 (C-4, C-4'), 75.3–73.0 (C-2, 3, 5), 60.2 (C-6).

Cellulose phosphate, heterogeneous synthesis in isopropanol (adapted from [351])

10 g (62 mmol) cellulose is added to a mixture of 232 g 85% H_3PO_4, 128 g (0.902 mol) P_2O_5 and 120 g isopropanol at 25 °C. After 72 h, the cellulose phosphate is isolated from the reaction mixture by centrifugation and washed with isopropanol. After stirring the product for 20 min in a mixture of 175 g methanol, 75 g water and 15 g sodium acetate, the product is filtered off, washed three times with 500 g methanol/water (3/1, w/w) and dried. DS 0.26. Elemental analysis: 4.5% P.

Starch phosphate, heterogeneous synthesis of cross-linked starch phosphate (adapted from [523])

2.5 g (15.4 mmol) starch and 15.0 ml Py are mixed and 5 ml Py are distilled off. Subsequently, 15.0 ml Py and 1.9 ml (20.8 mmol) $POCl_3$ are added at RT. The mixture is heated for 3 h at 70 °C. After cooling to RT, the product is filtered off and washed with water. NaOH (5.8%) was added to the slurry of the polymer in water until pH 8.0 was reached in order to convert the acid form to the sodium salt. The product was washed and dried. DS 0.92.

Cellulose nitrate, homogeneous synthesis in N_2O_4/DMF (adapted from [369])

N_2O_4 is bubbled into DMF until 3% is absorbed. 100 ml of this solution is added to 1.0 g (6.2 mmol) cellulose and the polymer dissolves within 15 min at RT. The solution is stirred for 3 h at 70 °C and subsequently cooled to RT. The product is precipitated in three volumes methanol under stirring, filtered off, washed with methanol free of acid, with acetone, and finally dried under vacuum at 50 °C. The polymer (DS 0.55) is water soluble. FTIR (KBr): 1660 ν (NO_2) cm^{-1}.

Cellulose nitrate, nitration with HNO_3/H_3PO_4 (adapted from [366])

The nitrating acid mixture is prepared by cautiously adding 1.6 g P_2O_5 to 40.0 g cold 90% nitric acid, stirring at 0 °C. After a few hours, the acid mixture is filtered through glass wool and allowed to warm to RT. 1.0 g (6.2 mmol) cellulose is added and allowed to react for 20 min and swirled every 5 min. The product is filtered off, poured into cold (10 °C) distilled water, stirred for a moment, and neutralised with powdered Na_2CO_3. The product is washed three times with water, boiled for 20 min in distilled water, filtered, soaked for 10 min in 50 ml methanol, filtered off, and dried at 50 °C. DS 2.9–3.0.

Cellulose trinitrate, polymer analogues synthesis (adapted from [368])

20 ml white fuming HNO_3 and 20 ml dichloromethane are mixed in a beaker and cooled to 0 °C. 0.75 g (4.6 mmol) dry cellulose is added and the mixture maintained at 0 °C for 30 min, with occasional stirring, in the case of hydrolytically degraded cellulose powder, the reaction time being extended to 90 min for high-DP cotton linters. The fibrous cellulose trinitrate is filtered off using a coarse sintered glass filtering crucible, washed three times with CH_2Cl_2, three times with methanol, once with water of 40–50 °C, and finally with methanol. The cellulose trinitrate is dried under vacuum at 20 °C. DS 3.0.

Perpropionylation of cellulose furoate (adapted from [234])

To 0.3 g cellulose furoate is added a mixture of 6 ml Py, 6 ml propionic acid anhydride and 50 mg DMAP. After 24 h at 80 °C, the reaction mixture is cooled to RT and precipitated in 50 ml ethanol. For purification, the isolated product is reprecipitated from chloroform into 50 ml ethanol, filtered off, washed with ethanol and dried under vacuum at RT. $DS_{Furoate}$ 1.49, $DS_{Propionyl}$ 1.50. FTIR (KBr): no ν (OH), 2910, 2854 ν (CH), 1758, 1737 ν (C=O_{Ester}) cm^{-1}. ^{13}C NMR (DMSO-d_6): δ (ppm) = 173.9–173.1 (C=$O_{Propionyl}$), 157.3 (C=$O_{Furoate}$), 147.5–112.2 (furan ring), 100.6–62.6 (modified RU), 27.2 (CH_2-propionate), 9.4 (CH_3-propionate). ^1H NMR ($CDCl_3$): δ (ppm) = 7.56 (H-11), 7.20 (H-9), 6.50 (H-10), 5.00 (H-3), 4.85 (H-2), 4.42 (H-1), 4.38 and 4.08 (H-6), 3.66, (H-4) 3.53 (H-5), 2.04 (CH_2-propionate), 1.06, 1.03 (CH_3-2, 3-propionate).

Permethylation of polysaccharide esters (adapted from [419])

5 mg of polysaccharide ester is dissolved in 2 ml trimethyl phosphate by ultrasonic agitation, followed by addition of 15 mol/mol free OH groups 2,6-di-tert-butyl-pyridine and 20 mol/mol free OH groups methyl triflate and ultrasonic treatment for 4 h (until 60 °C). The product is isolated by addition of chloroform and water and, after agitation, the chloroform phase is washed six times with water. The permethylated polysaccharides are purified through a Sephadex-LH 20 column with acetone as eluent. DS 3.0.

6-mono-O-(4-monomethoxy)trityl cellulose, homogeneous synthesis in DMAc/LiCl (adapted from [433])

1.0 g (6.2 mmol) cellulose and 100 ml DMAc are stirred for 30 min at 150 °C, to which at 100 °C 9.0 g dry LiCl is added. After stirring overnight, 2.2 ml dry Py is added over 10 min, followed by a solution of 5.7 g (18.5 mmol) 4-mono-methoxytrityl chloride in 50 ml DMAc over 20 min, under stirring at RT. The etherification was carried out at 70 °C for 24 h. The polymer is precipitated in methanol, filtered off, reprecipitated from DMF into methanol, and dried. DS 0.98.

2,6-di-O-Thexyldimethylsilyl cellulose in DMAc/LiCl (adapted from [446])

30.0 g (185 mmol) cellulose is suspended in 750 ml DMAc and the mixture stirred for 2 h at 120 °C. After cooling to 100 °C, 45.0 g anhydrous LiCl is added, followed by stirring at RT until complete dissolution of the polymer. 60.4 g (888 mmol) imidazole is added to this solution, followed by the addition of 154.4 ml (741 mmol) TDMSCl over 25 min. The reaction mixture is stirred at 100 °C. After 24 h, the mixture is poured into 1 l aqueous phosphate buffer solution (7.14 g K_2HPO_4 and 3.54 g KH_2PO_4 in 1 l distilled water). The precipitate is filtered off, washed with 5 l distilled water and 2.5 l ethanol, and dried under vacuum for 12 h while increasing the temperature up to 100 °C. DS 2.0. FTIR (KBr): 3500 ν (OH), 2961, 2873 ν (CH), 1467 ν (CH_2, CH_3), 1379 ν (CH_3), 1252 ν (Si−C), 1120, 1078, 1038 ν (C−O−C_{AGU}), 834, 777 ν (Si−C) cm^{-1}.

6-mono-O-Thexyldimethylsilyl cellulose in NMP/NH$_3$ (adapted from [524])

16.2 g (100 mmol) cellulose is suspended in 65 ml NMP and the mixture stirred at 80 °C for 1 h. After cooling to −25 °C, 80 ml NMP, saturated with ammonia, is added. The mixture is stirred for 1 h, and a solution of 39.25 ml (200 mmol) TDMSCl in 40 ml NMP is added drop-wise. After stirring for 45 min at −25 °C, the mixture is slowly warmed to 40 °C, allowed to stand overnight and then stirred for 6.5 h at 80 °C. The highly viscous solution is poured into 4 l of a buffer solution (pH 7). The polymer was filtered off, washed with water, and dried carefully at < 0.1 Torr over KOH and successively increasing the temperature from 25 to 80 °C.

For purification, it is dissolved in NMP, precipitated in buffer solution, washed and dried. DS 0.7.

Synthesis of 2-O-tosyl starch in DMAc/LiCl (adapted from [259])

To a 4.3% (w/v) solution of 2.5 g (15.4 mmol) starch in DMAc/LiCl is added a mixture of 7.4 ml (53.4 mmol) TEA and 5.0 ml DMAc, with stirring. After cooling to 8 °C, a solution of 5.9 g (30.8 mmol) TosCl in 7.4 ml DMAc is added drop-wise with stirring over 30 min. The homogeneous reaction mixture is additionally stirred for 24 h at 8 °C and subsequently slowly poured into 700 ml ice water. The precipitate is filtered off, carefully washed with 2.3 l distilled water and 250 ml ethanol, dissolved in 120 ml acetone and reprecipitated into 370 ml distilled water. After filtration and washing with ethanol, the product is dried for 8 h at 60 °C under vacuum. The product (DS 1.35) is soluble in acetone, DMSO, DMAc, DMF, THF and dioxane. FTIR (KBr): 3488 ν (OH), 3064 ν (C$-$H$_{Aromat}$), 2943 ν (CH), 1599, 1495, 1453 ν (C$-$C$_{Aromat}$), 1362 ν_{as}(SO$_2$), 1176 ν_s(SO$_2$), 811 δ (C$-$H$_{Aromat}$) cm^{-1}. ^{13}C NMR (DMSO-d_6): δ (ppm) = 21.1 (CH$_3$), 59.8$-$93.7 (starch backbone), 128.0$-$144.8 (C$-$H$_{Aromat}$). ^{1}H NMR (DMSO-d_6): δ (ppm) = 2.4 (CH$_3$), 3.4$-$5.6 (starch backbone), 7.4$-$7.8 (C$-$H$_{Aromat}$).

Azidothymidine-bound curdlan, homogeneous synthesis in DMSO (adapted from [490])

0.14 g (0.61 mol/mol AGU) 3$'$-azido-3$'$-deoxy-5$'$-O-succinyl-thymidine and 0.10 g DMAP are added to a solution of 0.10 g (0.62 mmol) curdlan in 3 ml DMSO. 0.30 g DCC is added gradually over 30 min. The mixture is stirred for 48 h at RT, followed by the addition of acetone until precipitation occurs. The precipitate is collected by centrifugation and washed 2 times with 30 ml acetone and 30 ml H$_2$O respectively. The white powdery azidothymidine-bound curdlan is obtained by freeze drying of the aqueous solution. DS 0.20. ^{13}C NMR (D$_2$O): δ (ppm) 172 (C=O), 64 (C-6$'$ and C-5$'$ of the dextran), 28 (CH$_3$).

Cellulose 3-(2-furyl)acrylcarboxylic acid ester, homogeneous synthesis in DMAc/LiCl and in situ activation with CDI (adapted from [234])

5.0 g (31.0 mmol) CDI is added to a solution of 4.3 g (31.0 mmol) 3-(2-furyl)-acrylcarboxylic acid in 20 m DMAc. The clear solution is stirred for 24 h at 40 °C and then added to a solution of 1 g (6.2 mmol) cellulose dissolved in DMAc/LiCl. The reaction mixture is stirred for 24 h at 60 °C and covered with aluminium foil. The product is precipitated into 250 ml ethanol, followed by filtration, and washing with 100 ml ethanol three times and drying at 45 °C under vacuum. DS 1.14. FTIR (KBr): 3490 ν (OH), 3132 ν(C$-$H$_{Furan}$), 2932 ν (C$-$H), 1711 ν (C=O$_{Ester}$), 1579 ν (furan ring), 1233 ν (C$-$O$-$C$_{Ester}$) cm^{-1}. ^{13}C NMR (DMSO-d_6): δ (ppm) = 165.3

(C=O), 150.1 (C-10), 145.8 (C-9, 13), 131.6 (C-8), 116.5 (C-12), 112.5 (C-11), 102.9 (C-1), 99.8 (C-1′), 72.3 (C-2), 73.9 (C-3), 80.2 (C-4), 76.3 (C-5), 63.1 (C-6$_s$), 60.2 (C-6). ^{1}H NMR (of the perpropionylated derivative dissolved in CDCl$_3$): δ (ppm) = 7.82 (H-11), 7.50 (H-8), 6.87 (H-9), 6.57 (H-10), 6.23 (H-7), 5.56 (H-1), 5.00 (H-3), 4.85 (H-2), 4.38 and 4.08 (H-6), 3.66, 3.63 (H-4, 5), 2.04 (CH$_2$-propionate), 0.77, 0.93 (CH$_3$-2, 3-propionate).

Cellulose 4′-carboxy-18-crown-6 ester, homogeneous synthesis and in situ activation with CDI (adapted from [234])

To a solution of 5 g (14.1 mmol) 4′-carboxybenzo-18-crown-6 in 20 ml DMAc is added 2.31 g (14.1 mmol) CDI in 20 ml DMAc, and the resulting mixture is stirred 24 h at RT. This mixture is added to a solution of 1.0 g (6.2 mmol) cellulose in DMAc/LiCl, and the homogeneous reaction mixture is stirred for 24 h at 70 °C under N$_2$. The product is precipitated into 250 ml ethanol, followed by filtration and washing with 100 ml ethanol three times, and drying at 45 °C under vacuum. DS 0.4. FTIR (KBr): 3410 ν (OH), 1714 ν (C=O$_{Ester}$), 1599 ν (ring stretch, benzene ring) cm^{-1}. ^{13}C NMR (DMSO-d_6): δ (ppm) = 165.8 (C=O), 113.3, 114.8, 148.5, 153.3 (C$_{Aromat}$), 103.2 (C-1), 80.5 (C-4), 73.2–77.2 (C-2, 3, 5), 69.3, 70.4 (crown ether CH$_2$), 62.4 (C-6$_s$), 60.9 (C-6). ^{1}H NMR (of the perpropionate in CDCl$_3$): δ (ppm) = 6.89–7.55 (H aromatic, H-9, H-10, H-13), 5.00 (H-3), 4.85 (H-2), 4.38 and 4.08 (H-6), 3.66, 3.63 (H-4, 5), 4.14 (H-14), 3.87 (H-15), 3.44–3.69 (H-16–18), 2.04 (CH$_2$-propionate), 0.77–1.17 (CH$_3$-propionate).

Enzymatic hydrolysis of starch (adapted from [130])

1.7 l water (pro analysis, FLUKA) is boiled under reflux for 1 h, cooled to RT, and 60 g (370 mmol) starch (Amioca, National Starch) is added slowly with vigorous stirring and heated under reflux for 1 h. The solution is stirred overnight at RT and then thermostated to 37 ± 0.1 °C. 0.5 g CaCl$_2$ and 0.005 g (130 U/l) amylase are added, and after 130 min the enzyme is thermally denaturised by heating for 20 min at 110 °C. Most of water is removed by evaporation, followed by dialysis and finally freeze drying. M_w 4.60 × 10^4 g/mol.

Degradation of chitosan (adapted from [70])

A solution of 100 mg chitosan in 10 ml 0.07 M HCl is gently shaken overnight at RT, followed by addition of 3 mg NaNO$_2$, storing for 4 h at RT and then lyophilisation.

DS determination of polysaccharide esters by saponification (adapted from [525])

0.5 g of the dry polysaccharide ester is swollen in 25 ml acetone/water mixture (1:1, v/v) for 24 h at RT. 12.5 ml 1 M KOH in ethanol is added and the mixture stirred for 24 h at RT. The excess alkali is titrated with 0.5 M aqueous HCl, using phenolphthalein as indicator. An excess of 2 ml 0.5 M HCl is added and back-titrated after 24 h with 0.5 M NaOH. From the titration results, the total amount of alkali consumed for saponification of the ester groups is obtained (in moles). Calculation of DS is performed according to Eq. (12.1).

$$DS_{Ester} = \frac{n\,KOH \cdot M_r\,(RU)}{m_s - M_r\,(RCO) \cdot n\,KOH} \tag{12.1}$$

$n\,KOH$ = total amount in moles of alkali consumed for saponification of the ester
$M_r\,(RU)$ = molar mass of the RU (e.g. 162 g/mol for glucans)
m_S = weight of the sample in g
$M_r\,(RCO)$ = molar mass of the ester group introduced

References

1. Bogan RT, Brewer RJ (1985) Cellulose esters, organic, survey. In: Mark HF, Bikales NM, Overberger CG, Menges G, Kroshwitz JI (eds) Encyclopedia of Polymer Science and Engineering. Wiley, New York, pp 158
2. Serad GA (1985) Cellulose esters, organic, fibres In: Mark HF, Bikales NM, Overberger CG, Menges G, Kroshwitz JI (eds) Encyclopedia of Polymer Science and Engineering. Wiley, New York, pp 200
3. Müller F, Leuschke C (1996) Organic cellulose esters, thermoplastic molding compounds. In: Bottenbruch L (ed) Engineering Thermoplastics: Polycarbonates-Polyacetals-Polyesters-Cellulose Esters. Hanser, Munich, pp 385
4. Kennedy JF, Griffiths AJ, Philp K, Stevenson DL, Kambanis O, Gray CJ (1989) Carbohydr Polym 11:1
5. Pelletier S, Hubert P, Lapicque F, Payan E, Dellacherie E (2000) Carbohydr Polym 43:343
6. Klemm D, Schmauder H-P, Heinze T (2002) Cellulose. In: De Baets S, Vandamme EJ, Steinbüchel A (eds) Biopolymers. Polysaccharides II, vol 6. Wiley, Weinheim, pp 275
7. Nakata M, Kawaguchi T, Kodama Y, Konno A (1998) Polymer 39:1475
8. Giavasis I, Harvey LM, McNeil B (2002) Scleroglucan. In: De Baets S, Vandamme EJ, Steinbüchel A (eds) Biopolymers. Polysaccharides II, vol 6. Wiley, Weinheim, pp 37
9. Rau U (2002) Schizophyllan. In: De Baets S, Vandamme EJ, Steinbüchel A (eds) Biopolymers. Polysaccharides II, vol 6. Wiley, Weinheim, pp 61
10. Misaki A, Kishida E, Kakuta M, Tabata K (1993) Antitumor fungal $(1-3)$-β-D-glucans: structural diversity and effects of chemical modification. In: Yalpani M (ed) Carbohydrates and Carbohydrate Polymers. ALT Press, Mount Prospect, IL, pp 116
11. Huynh R, Chaubet F, Jozefonvicz J (1998) Angew Makromol Chem 254:61
12. Shingel KI (2004) Carbohydr Res 339:447
13. Shogren RL (1998) Starch: properties and material applications. In: Kaplan DL (ed) Biopolymers from Renewable Resources. Springer, Berlin, Heidelberg, New York, pp 30
14. Ebringerová A, Heinze T (2000) Macromol Rapid Commun 21:542
15. Maier H, Anderson M, Karl C, Magnuson K, Whistler RL (1993) Guar, locust bean, tara, and fenugreek gums. In: BeMiller JN, Whister RL (eds) Industrial Gums. Polysaccharides and Their Derivatives, 3rd edn. Academic Press, San Diego, pp 181
16. Franck A, De Leenheer L (2002) Inulin. In: De Baets S, Vandamme EJ, Steinbüchel A (eds) Biopolymers. Polysaccharides II, vol 6. Wiley, Weinheim, pp 439
17. Roberts GAF (1992) Chitin Chemistry. Macmillan, London, pp 185
18. Sabra W, Deckewer W-D (2005) Alginate – a polysaccharide of industrial interest and diverse biological functions. In: Dumitriu S (ed) Polysaccharides. Structural Diversity and Functional Versatility, 2nd edn. Marcel Dekker, New York, pp 515
19. Dumitriu S (2005) Polysaccharides. Structural Diversity and Functional Versatility, 2nd edn. Marcel Dekker, New York
20. Vandamme EJ, De Baets S, Steinbüchel A (2002) Biopolymers. Polysaccharides I and II, vols 5 and 6. Wiley, Weinheim
21. Kaplan DL (1998) Biopolymers from renewable resources. Springer, Berlin Heidelberg New York
22. Ebert G (1993) Biopolymere: Struktur und Eigenschaften. Teubner, Stuttgart
23. Klemm D, Philipp B, Heinze T, Heinze U, Wagenknecht W (1998) Comprehensive Cellulose Chemistry, vols I and II. Wiley, New York
24. O'Sullivan AC (1997) Cellulose 4:173
25. Lai Y-Z (1996) Reactivity and accessibility of cellulose, hemicelluloses, and lignins. In: Hon DN-S (ed) Chemical Modification of Lignocellulosic Materials. Marcel Dekker, New York, pp 35

26. Philipp B (1993) J Macromol Sci Pure Appl Chem A30:703
27. Heinze T, Dicke R, Koschella A, Kull AH, Klohr E-A, Koch W (2000) Macromol Chem Phys 201:627
28. Tarchevsky IA, Marchenko GN (1991) Cellulose: Biosynthesis and Structure. Springer, Berlin Heidelberg New York
29. Lee I-Y (2002) Curdlan. In: Vandamme EJ, De Baets S, Steinbüchel A (eds) Biopolymers. Polysaccharides I, vol 5. Wiley, Weinheim, pp 135
30. Rau U, Mueller RJ, Cordes K, Klein J (1990) Bioprocess Eng 5:89
31. Taylor C, Cheetham NWH, Walker GJ (1985) Carbohydr Res 137:1
32. Leathers TD (2002) Dextran. In: Vandamme EJ, De Baets S, Steinbüchel A (eds) Biopolymers. Polysaccharides I, vol 5. Wiley, Weinheim, pp 299
33. Vandamme EJ, Bruggeman G, De Baets S, Vanhooren PT (1996) Agro-Food-Ind Hi-Tech 7:21
34. Leathers TD (2002) Pullulan. In: De Baets S, Vandamme EJ, Steinbüchel A (eds) Biopolymers. Polysaccharides II, vol 6. Wiley, Weinheim, pp 1
35. Dais P, Vlachou S, Taravel FR (2001) Biomacromolecules 2:1137
36. Lindblad MS, Albertsson A-C (2005) Chemical modification of hemicelluloses and gums. In: Dumitriu S (ed) Polysaccharides. Structural Diversity and Functional Versatility, 2nd edn. Marcel Dekker, New York, pp 491
37. Stevens CV, Meriggi A, Booten K (2001) Biomacromolecules 2:1
38. Peter MG (2002) Chitin and chitosan from animal sources. In: De Baets S, Vandamme EJ, Steinbüchel A (eds) Biopolymers. Polysaccharide II, vol 6. Wiley, Weinheim, pp 481
39. Valla S, Ertesvåg H, Skjåk-Bræk G (1996) Carbohydr Eur 14:14
40. Dische Z (1962) Color reactions of carbohydrates. In: Whistler RL, Wolfrom ML (eds) The Carbohydrates: Chemistry and Biochemistry, vol IB. Academic Press, New York, pp 475
41. Dubois M, Gilles KA, Hamilton JK, Rebers PA, Smith F (1956) Anal Chem 28:350
42. Blumenkrantz N, Asboe-Hansen G (1973) Anal Chem 54:484
43. Lever M (1972) Anal Chem 47:273
44. Updegraff DM (1969) Anal Chem 32:420
45. Jiang X, Chen L, Zhong W (2003) Carbohydr Polym 54:457
46. Kostromin AI, Abdullin IF, Ibeneeva RM (1982) Zavod Lab 48:16
47. Hsiao H-Y, Tsai C-C, Chen S, Hsieh B-C, Chen RLC (2004) Macromol Biosci 4:919
48. Fengel D, Ludwig M (1991) Papier 45:45
49. Mackie W (1971) Carbohydr Res 20:413
50. Jayme G, Tio PK (1968) Papier 22:322
51. Kath F, Kulicke W-M (1999) Angew Makromol Chem 268:69
52. Mais U, Sixta H (2004) Characterization of alkali-soluble hemicelluloses of hardwood dissolving pulps. In: Gatenholm P, Tenkanen M (eds) Hemicelluloses: Science and Technology. ACS Symposium Series 864, American Chemical Society, Washington, DC, pp 94
53. Zou Y, Khor E (2005) Biomacromolecules 6:80
54. Kulicke W-M, Otto M, Baar A (1993) Makromol Chem 194:751
55. Kim Y-T, Kim E-H, Cheong C, Williams DL, Kim C-W, Lim S-T (2000) Carbohydr Res 328:331
56. Bociek SM, Izzard MJ, Morrison A, Welti D (1981) Carbohydr Res 93:279
57. McIntyre DD, Vogel HJ (1993) Starch/Staerke 45:406
58. Gagnaire D, Vignon M (1977) Makromol Chem 178:2321
59. Grasdalen H, Larsen B, Smidsrød O (1981) Carbohydr Res 89:179
60. Høidal HK, Ertesvåg H, Skjåk-Bræk G, Stokke BT, Valla S (1999) J Biol Chem 274:12316
61. Atalla RH, VanderHart DL (1984) Science (Washington, DC) 223:283
62. Isogai A, Usuda M, Kato T, Uryu T, Atalla RH (1989) Macromolecules 22:3168
63. Nehls I, Wagenknecht W, Philipp B, Stscherbina D (1994) Prog Polym Sci 19:29
64. Hasegawa M, Isogai A, Onabe F, Usada M (1992) J Appl Polym Sci 45:1857
65. Flugge LA, Blank JT, Petillo PA (1999) J Am Chem Soc 121:7228
66. Cheetham NWH, Fiala-Beer E (1990) Carbohydr Polym 14:149
67. Tsai PK, Frevert J, Ballou CE (1984) J Biol Chem 259:3805
68. Grasdalen H, Larsen B, Smidsrød O (1979) Carbohydr Res 68:23
69. Grasdalen H (1983) Carbohydr Res 118:255
70. Vårum KM, Anthonsen MW, Grasdalen H, Smidsrød O (1991) Carbohydr Res 211:17

71. Kiemle DJ, Stipanovic AJ, Mayo KE (2004) Proton NMR methods in the compositional characterization of polysaccharides. In: Gatenholm P, Tenkanen M (eds) Hemicelluloses: Science and Technology. ACS Symposium Series 864. American Chemical Society, Washington, DC, pp 122
72. Hounsell EF (1995) Prog Nucl Magn Resonan Spectrosc 27:445
73. Skjåk-Bræk G, Grasdalen H, Larsen B (1986) Carbohydr Res 154:239
74. Grasdalen H, Kvam BJ (1986) Macromolecules 19:1913
75. Knutson CA, Jeanes A (1968) Anal Biochem 24:482
76. Grasdalen H, Larsen B, Smidsrød O (1979) Uronic acid sequence in alginates by carbon-13 nuclear magnetic resonance. In: Proc Int Seaweed Symp, 1977, vol 9, pp 309
77. Honda S, Suzuki S, Takahashi M, Kakehi K, Ganno S (1983) Anal Biochem 134:34
78. Cheetham NWH, Sirimanne P (1983) Carbohydr Res 112:1
79. Annison G, Cheetham NWH, Couperwhite I (1983) J Chromatogr 264:137
80. Franz G (1991) Polysaccharide. Springer, Berlin Heidelberg New York
81. Hakomori S (1964) J Biochem 55:205
82. D'Ambra AJ, Rice MJ, Zeller SG, Gruber PR, Gray GR (1988) Carbohydr Res 177:111
83. Biermann J, McGinnis GD (1989) Analysis of Carbohydrates by GLC and MS. CRC Press, Boca Raton
84. Dahlman OB, Jacobs A, Nordstroem M (2004) Characterization of hemi-celluloses from wood employing matrix-assisted laser desorption/ionization time-of-flight mass spectrometry. In: Gatenholm P, Tenkanen M (eds) Hemicelluloses: Science and Technology. ACS Symposium Series 864. American Chemical Society, Washington, DC, pp 80
85. Tanghe LJ, Brewer RJ (1955) Anal Chem 40:350
86. March J (1992) Advanced Organic Chemistry: Reactions, Mechanisms, and Structure, 4th edn. Wiley, New York, pp 170
87. Jarowenko W (1987) Acetylated starch and miscellaneous organic esters. In: Wurzburg OB (ed) Modified Starches: Properties and Uses. CRC Press, Boca Raton, pp 55
88. Doyle S, Pethrick RA, Harris RK, Lane JM, Packer KJ, Heatley F (1986) Polymer 27:19
89. Miyamoto T, Sato Y, Shibata T, Tanahashi M, Inagaki H (1985) J Polym Sci Polym Chem Ed 23:1373
90. Tedder JM (1955) Chem Rev (Washington, DC) 55:787
91. Hamalainen C, Wade RH, Buras EM Jr (1957) Text Res J 27:168
92. Novak LJ, Tyree JT (1960) US 2954372 CAN 55:50844
93. Iwata T, Fukushima A, Okamura K, Azuma J (1997) J Appl Polym Sci 65:1511
94. Mueller F (1985) Papier (Bingen, Germany) 39:591
95. Morooka T, Norimoto M, Yamada T, Shiraishi N (1984) J Appl Polym Sci 29:3981
96. Bourne EJ, Stacey M, Tatlow JC, Tedder JM (1949) J Chem Soc Abstr 2976
97. Sun RC, Sun XF, Tomkinson J (2004) Hemicelluloses and their derivatives. In: Gatenholm P, Tenkanen M (eds) Hemicelluloses: Science and Technology. ACS Symposium Series 864. American Chemical Society, Washington, DC, pp 2
98. Sun X-F, Sun R-C, Zhao L, Sun J-X (2004) J Appl Polym Sci 92:53
99. Edgar KJ, Pecorini TJ, Glasser WG (1998) Long-chain cellulose esters: preparation, properties, and perspective. In: Heinze T, Glasser WG (eds) Cellulose Derivatives. Modification, Characterization and Nanostructures. ACS Symposium Series 688. American Chemical Society, Washington, DC, pp 38
100. Wurzburg OB (1964) Acetylation. In: Whistler RL (ed) Methods in Carbohydrate Chemistry, vol 4. Academic Press, New York, pp 286
101. Höfler G, Steglich W, Vorbrüggen H (1978) Angew Chem 90:602
102. Tezuka Y (1998) Carbohydr Res 305:155
103. Kurita K, Ishii S, Tomita K, Nishimura S-I, Shimoda K (1994) J Polym Sci Polym Chem Ed 32:1027
104. Hiatt GD, Mench JW, Emerson J (1956) US 2759925 CAN 50:79767
105. Wade RH, Reeves WA (1964) Text Res J 34:836
106. Aburto J, Alric I, Thiebaud S, Borredon E, Bikiaris D, Prinos J, Panayiotou C (1999) J Appl Polym Sci 74:1440
107. Ehrhardt S, Begli AH, Kunz M, Scheiwe L (1997) EP 792888 CAN 127:249640
108. Sircar AK, Stanonis DJ, Conrad CM (1967) J Appl Polym Sci 11:1683
109. Malm CJ, Mench JW, Kendall DL (1951) J Ind Eng Chem (Seoul) 43:684
110. Novac LJ, Tyree JT (1956) US 2734005 CAN 50:46838
111. Malm CJ, Mench JW, Kendall DL, Hiatt GD (1951) J Ind Eng Chem (Seoul) 43:688
112. Gros AT, Feuge RO (1962) J Am Oil Chem Soc 39:19

113. Sagar AD, Merrill EW (1995) J Appl Polym Sci 58:1647
114. Teng J, Rha C (1972) US 3666492 CAN 77:60348
115. Bader H, Rafler G, Lang J, Lindhauer M, Klaas MR, Funke U, Warwel S (1998) EP 859012 CA 129:204389
116. Riemschneider R, Siekfeld J (1964) Monatsh Chem 95:194
117. Peltonen S, Harju K (1994) WO 9422919 CAN 123:86469
118. Wang P, Tao BY (1994) J Appl Polym Sci 52:755
119. Wang P, Tao BY (1999) Characterization of plasticized and mixed long-chain fatty cellulose esters. In: Imam SH, Green RV, Zaidi BR (eds) Biopolymers: Utilizing Nature's Advanced Materials. ACS Symposium Series 723, pp 77
120. Shogren RL (2003) Carbohydr Polym 52:319
121. Hirata Y, Aoki M, Kobatake H, Yamamoto H (1999) Biomaterials 20:303
122. Heinze T, Glasser WG (1998) The role of novel solvents and solution complexes for the preparation of highly engineered cellulose derivatives. In: Heinze T, Glasser WG (eds) Cellulose Derivatives. Modification, Characterisation and Nanostructures. ACS Symposium Series 688, pp 2
123. Husemann E, Siefert E (1969) Makromol Chem 128:288
124. Wu J, Zhang J, Zhang H, He J, Ren Q, Guo M (2004) Biomacromolecules 5:266
125. Klohr EA, Koch W, Klemm D, Dicke R (2000) DE 19951734 CAN 133:224521
126. Ibrahim AA, Nada AMA, Hagemann U, El Seoud OA (1996) Holzforschung 50:221
127. Heinze T, Liebert T, Pfeiffer KS, Hussain MA (2003) Cellulose 10:283
128. Takaragi A, Minoda M, Miyamoto T, Liu HQ, Zhang LN (1999) Cellulose 6:93
129. Ciacco GT, Liebert TF, Frollini E, Heinze TJ (2003) Cellulose (Dordrecht, Netherlands) 10:125
130. Förster H, Asskali F, Nitsch E (1991) DE 4123000 CAN 118:175891
131. Mark AM, Mehltretter CL (1972) Starch/Staerke 24:73
132. Shogren RL (1996) Carbohydr Polym 29:57
133. Billmers RL, Tessler MM (1994) AU 648956 B1 CAN 121:207918
134. Brockway CE (1965) J Polym Sci Part A. Polym Chem 3:1031
135. Parmerter MS (1971) US 3620913 CAN 76:73995
136. Fang JM, Fowler PA, Sayers C, Williams PA (2004) Carbohydr Polym 55:283
137. Lipparini L, Garutti MA (1967) Quad Merceol 5:35
138. Tessler MM (1973) DE 2230884 CAN 78:99447
139. Sun RC, Sun XF, Bing X (2002) J Appl Polym Sci 83:757
140. Fischer S, Voigt W, Fischer K (1999) Cellulose (Dordrecht, Netherlands) 6:213
141. Leipner H, Fischer S, Brendler E, Voigt W (2000) Macromol Chem Phys 201:2041
142. Fischer S, Leipner H, Brendler E, Voigt W, Fischer K (1999) ACS Symp Ser 737:143
143. Narayan R, Bloembergen ST, Lathia A (1995) WO 9504083 CAN 123:116083
144. Bruneel D, Schacht E (1994) Polymer 35:2656
145. Vermeersch J, Schacht E (1985) Bull Soc Chim Belg 94:287
146. Bauer KH, Reinhart T (1996) DE 4433101 CAN 124:292795
147. Reinisch G, Radics U, Roatsch B (1995) Angew Makromol Chem 233:113
148. Shimooozono T, Shiraishi N (1997) JP 09031103 CAN 126:226739
149. Ferruti P, Tanzi MC, Vaccaroni F (1979) Makromol Chem 180:375
150. Philipp B (1990) Polym News 15:170
151. Deus C, Friebolin H, Siefert E (1991) Makromol Chem 192:75
152. Swatloski RP, Spear SK, Holbrey JD, Rogers RD (2002) J Am Chem Soc 124:4974
153. Heinze T, Barthel S, Schwikal K (2005) Macromol Biosci 5:520
154. Sanchez-Chavez M, Arranz F, Diaz C (1989) Makromol Chem 190:2391
155. Ramirez JC, Sanchez-Chavez M, Arranz F (1994) Polymer 35:2651
156. Kim S-H, Chu C-C (2000) J Biomed Mater Res 49:517
157. Sun RC, Fang JM, Tomkinson J, Hill CAS (1999) J Wood Chem Technol 19:287
158. McCormick CL, Lichatowich DK, Pelezo JA, Anderson KW (1980) ACS Symp Ser 121:371
159. El-Kafrawy A (1982) J Appl Polym Sci 27:2435
160. Spange S, Reuter A, Vilsmeier E, Heinze T, Keutel D, Linert W (1998) J Polym Sci Polym Chem Ed 36:1945
161. Vincendon M (1985) Makromol Chem 186:1787
162. McCormick CL, Callais PA (1987) Polymer 28:2317
163. El Seoud OA, Marson GA, Ciacco GT, Frollini E (2000) Macromol Chem Phys 201:882

164. Regiani AM, Frollini E, Marson GA, Arantes GM, El Seoud OA (1999) J Polym Sci Polym Chem Ed 37:1357
165. Pawlowski WP, Sankar SS, Gilbert RD, Fornes RD (1987) J Polym Sci Polym Chem Ed 25:3355
166. Fang JM, Fowler PA, Tomkinson J, Hill CAS (2002) Carbohydr Polym 47:245
167. Terbojevich M, Cosani A, Carraro C, Torri G (1988) In: Skjak-Braek G, Anthonsen T, Sandford P (eds) Chitin and Chitosan. Elsevier, London, pp 407
168. Grote C, Heinze T (2005) Cellulose (Dordrecht, Netherlands) 12:435
169. Gräbner D, Liebert T, Heinze T (2002) Cellulose (Dordrecht, Netherlands) 9:193
170. Hon DN, Yan HJ (2001) J Appl Polym Sci 81:2649
171. McCormick CL, Lichatowich DK (1979) J Polym Sci Polym Lett Ed 17:497
172. Terbojevich M, Cosani A, Focher B, Gastaldi G, Wu W, Marsano E, Conio G (1999) Cellulose (Dordrecht, Netherlands) 6:71
173. Yoshida Y, Yanagisawa M, Isogai A, Suguri N, Sumikawa N (2005) Polymer 46:2548
174. Sharma RK, Fry JL, James L (1983) J Org Chem 48:2112
175. Sun H, DiMagno SG (2005) J Am Chem Soc 127:2050
176. Köhler S (2005) Neue Lösemittel für Cellulose DMSO/Ammoniumfluoride. Diploma Thesis, University of Jena, Germany
177. Zhang L, Zhang M, Zhou Q, Chen J, Zeng F (2000) Biosci Biotechnol Biochem 64:2172
178. Whistler RL, Roberts HJ (1959) J Am Chem Soc 81:4427
179. Deuel H, Solms J, Neukom H (1954) Chimia 8:64
180. Fujimoto T, Takahashi S, Tsuji M, Miyamoto T, Inagaki H (1986) J Polym Sci Part C. Polym Lett 24:495
181. Trapasso LE (1977) US 4011393 CAN 86:157395
182. Aburto J, Hamaili H, Mouysset-Baziard G, Senocq F, Alric I, Borredon E (1999) Starch/Staerke 51:302
183. Emelyanov YG, Grinspan DD, Kaputskii FN (1988) Khim Drev 1:23
184. Salin BN, Cemeris M, Mironov DP, Zatsepin AG (1991) Khim Drev 3:65
185. Schnabelrauch M, Vogt S, Klemm D, Nehls I, Philipp B (1992) Angew Makromol Chem 198:155
186. Philipp B, Wagenknecht W, Nehls I, Ludwig J, Schnabelrauch M, Kim HR, Klemm D (1990) Cellul Chem Technol 24:667
187. Liebert T, Klemm D, Heinze T (1996) J Macromol Sci Pure Appl Chem A33:613
188. Liebert T, Schnabelrauch M, Klemm D, Erler U (1994) Cellulose (Dordrecht, Netherlands) 1:249
189. Mansson P, Westfelt L (1980) Cellul Chem Technol 14:13
190. Seymour RB, Johnson EL (1978) J Polym Sci Polym Chem Ed 16:1
191. Clermont LP, Manery N (1974) J Appl Polym Sci 18:2773
192. Wagenknecht W, Nehls I, Philipp B (1993) Carbohydr Res 240:245
193. Johnson DC, Nicholson MD, Haigh FG (1976) J Appl Polym Sci Appl Polym Symp 28:931
194. Baker TJ, Schroeder LR, Johnson DC (1981) Cellul Chem Technol 15:311
195. Saikia CN, Dutta NN, Borah M (1993) Thermochim Acta 219:191
196. Shimizu Y, Hayashi J (1988) Sen'i Gakkaishi 44:451
197. Shimizu Y, Nakayama A, Hayashi J (1991) Cellul Chem Technol 25:275
198. Jandura P, Riedl B, Kokta BV (2000) Polym Degrad Stabil 70:387
199. Schaller J, Heinze T (2000) Macromol Chem Phys 201:1214
200. Brewster JW, Ciotti CJ (1955) J Am Chem Soc 77:6214
201. Sealey JE, Frazier CE, Samaranayake G, Glasser WG (2000) J Polym Sci Polym Phys Ed 38:486
202. Sanchez-Chaves M, Arranz F (1997) Polymer 38:2501
203. Vasil'ev AE, Khachatur'yan AA, Rozenberg GY (1971) Khim Prir Soedin 7:698
204. Sealey JE, Samaranayake G, Todd JG, Glasser WG (1996) J Polym Sci Polym Phys Ed 34:1613
205. Glasser WG, Samaranayake G, Dumay M, Dave V (1995) J Polym Sci Polym Phys Ed 33:2045
206. Glasser WG, Becker U, Todd JG (2000) Carbohydr Polym 42:393
207. Koschella A, Haucke G, Heinze T (1997) Polym Bull 39:597
208. Gradwell SE, Rennacker S, Esker AR, Heinze T, Gatenholm P, Vaca-Garcia C, Glasser W (2004) CR Biol 327:945
209. Harada A, Shintani A, Sugiyama H, Iwamoto O (1986) JP 61152701 CAN 105:228854
210. Haslam E (1980) Tetrahedron 36:2409
211. Fenselau AH, Moffatt JG (1966) J Am Chem Soc 88:1762
212. Bamford CH, Middleton IP, Al-Lamee KG (1986) Polymer 27:1981
213. Won CY, Chu CC (1998) J Appl Polym Sci 70:953

214. Kochetkov NK, Khachatur'yan AA, Vasil'ev AE, Rozenberg GY (1969) Khim Prir Soedin 5:427
215. Azhigirova MA, Vasil'ev AE, Gerasimovskaya LA, Khachatur'yan AA, Rozenberg GY (1977) Zh Obshch Khim 47:464
216. Nichifor M, Carpov A (1999) Eur Polym J 35:2125
217. Nichifor M, Stanciu MC, Zhu XX (2004) React Funct Polym 59:141
218. Jung SW, Jeong YI, Kim YH, Kim SH (2004) Arch Pharmacol Res 27:562
219. Glinel K, Huguet J, Muller G (1999) Polymer 40:7071
220. Mocanu G, Carpov A, Chapelle S, Merle L, Muller G (1995) Can J Chem 73:1933
221. Kim S, Chae SY, Na K, Kim SW, Bae YH (2003) Biomaterials 24:4843
222. Vermeersch J, Vandoorne F, Permentier D, Schacht E (1985) Bull Soc Chim Belg 94:591
223. Samaranayake G, Glasser WG (1993) Carbohydr Polym 22:1
224. Zhang ZB, McCormick CL (1997) J Appl Polym Sci 66:293
225. Williamson SL, McCormick CL (1998) J Macromol Sci Pure Appl Chem A35:1915
226. Deguchi S, Akiyoshi K, Sunamoto J (1994) Macromol Rapid Commun 15:705
227. Auzely-Velty R, Rinaudo M (2003) Int J Biol Macromol 31:123
228. Staab HA (1962) Angew Chem 74:407
229. Kol'tsova GN, Khachatur'yan AA, Doronina TN, Vasil'ev AE, Rozenberg GY (1972) Khim Prir Soedin 3:266
230. Williams AS, Taylor G (1992) Int J Pharm 83:233
231. van Dijk-Wolthuis WNE, Tsang SKY, Kettenes-Van Den JJ, Henninka WE (1997) Polymer 38:6235
232. Neumann U, Wiege B, Warwel S (2002) Starch/Staerke 54:449
233. Grote C (2004) Stärkeester – Neue Synthesewege und Produkte. Ph. D. Thesis, University of Jena
234. Liebert TF, Heinze T (2005) Biomacromolecules 6:333
235. Hon DN, Yan HJ (2001) J Appl Polym Sci 82:243
236. Zhu C, Wang D, Hu H (1984) Kexue Tongbao (foreign language edn) 29:707
237. Tanida F, Tojima T, Han SM, Nishi N, Tokura S, Sakairi N, Seino H, Hamada K (1998) Polymer 39:5261
238. Sun T, Lindsay JD (2004) US 6689378 B1 CAN 135:78461
239. Hussain MA, Liebert T, Heinze Th (2004) Polym News 29:14
240. Vaca-Garcia C, Borredon ME, Gaset A (2000) WO 2000050493 CAN 133:194826
241. Laletin AJ, Gal'braikh LS, Rogovin ZA (1968) Vysokomol Soedin Ser A 10:652
242. Rooney ML (1976) Polymer 17:555
243. Ferreira L, Gil MH, Dordick JS (2002) Biomaterials 23:3957
244. Ferreira L, Carvalho R, Gil MH, Dordick JS (2002) Biomacromolecules 3:333
245. Dicke R (2004) Cellulose (Dordrecht, Netherlands) 11:255
246. Chakraborty S, Sahoo B, Teraoka I, Miller LM, Gross RA (2005) Macromolecules 38:61
247. Levesque G, Lemee L (2002) FR 2816310 CAN 137:354616
248. Edgar KJ, Arnold KM, Blount WW, Lawniczak JE, Lowman DW (1995) Macromolecules 28:4122
249. Witzemann JS, Nottingham WD, Rector FD (1990) J Coat Technol 62:101
250. Witzemann JS, Nottingham WD (1991) J Org Chem 56:1713
251. Clemens RJ (1986) Chem Rev (Washington, DC) 86:241
252. Wentrup C, Heilmayer W, Kollenz G (1994) Synthesis 1219
253. McCormick CL, Dawsey TR (1990) Macromolecules 23:3606
254. Donabedian DH, McCarthy SP (1998) Macromolecules 31:1032
255. Ikeda I, Washino K, Maeda Y (2003) Sen'i Gakkaishi 59:110
256. McCormick CL, Callais PA (1986) Polym Prepr (Am Chem Soc, Div Polym Chem) 27:91
257. Rahn K, Diamantoglou M, Klemm D, Berghmans H, Heinze Th (1996) Angew Makromol Chem 238:143
258. Clode DM, Horton D (1971) Carbohydr Res 17:365
259. Heinze T, Talaba P, Heinze U (2000) Carbohydr Polym 42:411
260. Kurita K, Yoshino H, Yokota K, Ando M, Inoue S, Ishii S, Nishimura S-I (1992) Macromolecules 25:3786
261. Terbojevich M, Carraro C, Cosani A, Marsano E (1988) Carbohydr Res 180: 73
262. Ernst B, Tessmann D, Stoehlmacher P (1971) Wiss Z Univ Rostock Math-Naturwiss Reihe 20:623
263. Heinze T, Camacho JA, Haucke G (1996) Polym Bull 37:743
264. Wolfrom ML, Sowden JC, Metcalf EA (1941) J Am Chem Soc 63:1688
265. Roberts RW (1957) J Am Chem Soc 79:1175
266. Frazier CE, Glasser WG (1990) Polym Prepr (Am Chem Soc, Div Polym Chem) 31:634

267. Siegmund G (2000) Untersuchungen zur Reaktivität von Cellulosesulfonaten in nucleophilen Substitutionsreaktionen und Folgederivatisierungen. Ph. D. Thesis, University of Jena, Germany
268. Mocanu G, Constantin M, Carpov A (1996) Angew Makromol Chem 241:1
269. Biermann CJ, Narayan R (1986) Carbohydr Res 153:C1
270. Tsukada Y, Nemori R (1990) JP 02129236 CAN 113:174352
271. Siegmund G, Klemm D (2002) Polym News 27:84
272. McCormick CL, Dawsey TR, Newman JK (1990) Carbohydr Res 208:183
273. Dicke R, Rahn K, Haack V, Heinze T (2001) Carbohydr Polym 45:43
274. Kurita K, Inoue S, Yamamura K, Yoshino H, Ishii S, Nishimura S (1992) Macromolecules 25:3791
275. Kurita K, Hashimoto S, Yoshino H, Ishii S, Nishimura S (1996) Macromolecules 29:1939
276. Morita Y, Sugahara Y, Takahashi A, Ibonai M (1994) Eur Polym J 30:1231
277. Ilieva NI, Gal'braikh LS (1975) Studies in nucleophilic substitution of cellulose benzenesulfonates. Gromov VS (ed) Tezisy Dokl – Vses Konf Khim Fiz Tsellyul 1st, 1, pp 64; Ilieva NI, Gal'braikh LS, Rogovin ZA (1975) Koksnes Kim 6:13; Ilieva NI, Gal'braikh LS, Rogovin ZA (1976) Cellul Chem Technol 10:547; Ilieva NI, Gal'braikh LS, Kolokolkina NV, Kolosova TE, Zhbankov RG (1976) Koksnes Kim 3:28
278. Snezhko VA, Khomyakov KP, Komar VP, Osipova VV, Virnik AD, Zhbankov RG, Rogovin ZA (1975) Zh Prikl Khim (S Peterburg) 48:1540
279. Kolova AF, Komar VP, Skornyakov IV, Virnik AD, Zhbanov RG, Rogovin ZA (1978) Cellul Chem Technol 12:553
280. Carson JF, Maclay WD (1948) J Am Chem Soc 70:2220
281. Boeykens S, Vazquez C, Temprano N (2003) Spectrochim Acta Part B 58B:2169
282. Boeykens SP, Vazquez C, Temprano N, Rosen M (2004) Carbohydr Polym 55:129
283. Berlin P, Tiller J, Klemm D (1997) WO 9725353 CA 127:150332
284. Yalpani M (1988) Polysaccharides. Syntheses, Modifications and Structure/Property Relations. Elsevier, Amsterdam, pp 8
285. Woodings CR (1993) Regenerated cellulosics. In: Kroschwitz JI, Howe-Grant M (eds) Encyclopedia of Chemical Technology, 4th edn, vol 10. Wiley, New York, pp 696
286. Schweiger RG (1971) DE 2120964 CAN 76:128029
287. Wagenknecht W, Nehls I, Philipp B (1992) Carbohydr Res 237:211
288. Fransson L-A (1985) Mammalian glycosaminoglycans. In: Aspinall GO (ed) The Polysaccharides. Academic Press, New York, pp 338
289. Nakano T, Dixon WT, Ozimek L (2002) Proteoglycans (glucosaminoglycans/mucopolysaccharides). In: De Baets S, Vandamme EJ, Steinbüchel A (eds) Biopolymers. Polysaccharides II, vol 6. Wiley, Weinheim, pp 575
290. Gallagher JT, Lyon M, Steward WP (1986) Biochem J 236:313
291. Rodén L (1968) The Protein-carbohydrate linkages of acid mucopolysaccharides. In: Quintarelli G (ed) Chemical Physiology of Mucopolysaccharides, vol date 1965, Churchill, London, pp 17
292. Malmstroem A, Aaberg L (1982) Biochem J 201:489
293. Koumoto K, Umeda M, Numata M, Matsumoto T, Sakurai K, Kunitake T, Shinkai S (2004) Carbohydr Res 339:161
294. Whistler RL, Spencer WW (1961) Arch Biochem Biophys 95:36
295. Osawa Z, Morota T, Hatanaka K, Akaike T, Matsuzaki K, Nakashima H, Yamamoto N, Suzuki E, Miyano H, Mimura T, Kaneko Y (1993) Carbohydr Polym 21:283
296. Yoshida T, Yasuda Y, Mimura T, Kaneko Y, Nakashima H, Yamamoto N, Uryu T (1995) Carbohydr Res 276:425
297. Chaidedgumjorn A, Toyoda H, Woo ER, Lee KB, Kim YS, Toida T, Imanari T (2002) Carbohydr Res 337:925
298. Wang Y-J, Yao S-J, Guan Y-X, Wu T-X, Kennedy JF (2005) Carbohydr Polym 59:93
299. Ronghua H, Yumin D, Jianhong Y (2003) Carbohydr Polym 52:19
300. Bischoff KH, Dautzenberg H (1975) DD 112456 CAN 84:107351
301. Roberts HJ (1967) Starch derivatives. In: Whistler RL, Paschall EF (eds) Starch: Chemistry and Technology, vol 2. Academic Press, New York, pp 293
302. Zurawski P, Skrzypczak W, Wrocinski T, Kaczmarek F (1976) PL 87000 CAN 88:24492
303. Koerdel K, Schierbaum F (1975) DD 117078 CAN 85:7578
304. Schierbaum F, Koerdel K (1978) Reaction of starch with chlorosulfonic acid-formamide reagent. Carbohydrate Sulfates, ACS Symp Series 77. American Chemical Society, pp 173

305. Paschall EF (1956) US 2775586 CAN 51:23928
306. Martin I, Wurzburg OB (1958) US 2857377 CAN 53:39631
307. Whistler RL, Spencer WW (1964) Sulfation. Triethylamin-sulfur trioxide complex. In: Whister RL (ed) Methods in Carbohydrate Chemistry, vol 4. Academic Press, New York, pp 297
308. Smith HE, Russel CR, Rist CE (1963) Cereal Chem 40:282
309. Whistler RL (1970) US 3507855
310. Guisely KB, Whitehouse PA (1973) US 3720659 CAN 79:20613
311. Kucerova M, Pasteka M (1975) Chem Zvesti 29:697
312. Kerr RW, Paschall EF, Minkema WmH (1961) US 2967178 CAN 55:39503
313. Tessler MM (1978) US 4093798 CAN 89:165292
314. Yao S (2000) Chem Eng J (Lausanne) 78:199
315. Klemm D, Philip B, Heinze T, Heinze U, Wagenknecht W (1998) Comprehensive Cellulose Chemistry, vol 2. Wiley, New York, pp 117
316. Schweiger RG (1974) Tappi 57:86
317. Wagenknecht W, Nehls I, Stein A, Klemm D, Philipp B (1992) Acta Polym 43:266
318. Tester RF, Karkalas J (2002) Starch. In: De Baets S, Vandamme EJ, Steinbüchel A (eds) Biopolymers. Polysaccharides II. Wiley, New York, pp 381
319. Blennow A, Bay-Smidt AM, Wischmann B, Olsen CE, Moller BL (1998) Carbohydr Res 307:45
320. Blennow A, Nielsen TH, Tom H, Bausngaard L, Mikkelsen R, Engelsen SB (2002) Trends Plant Sci 7:445
321. Hamilton RM, Paschall EF (1967) Production and uses of starch phosphates. In: Whistler RL, Paschall EF (eds) Starch: Chemistry and Technology, vol 2. Academic Press, New York, pp 351
322. Paschall EF (1964) Phosphation with inorganic phosphate salts. In: Whistler RL (ed) Methods in Carbohydrate Chemistry, vol 4. Academic Press, New York, pp 294
323. Rubens RW, Patel JK, Wurzburg OB, Jarowenko W (1980) US 4216310 CAN 93:188098
324. Tessler MM (1974) US 3838149 CAN 82:74773
325. Tessler MM (1974) US 3842071 CAN 82:74775
326. Tessler MM, Rutenberg MW (1973) US 3720662 CAN 79:20608
327. Verbanac F, Moser KB (1971) US 3553194 CAN 74:100820
328. Tessler MM (1976) US 3969341 CAN 85:14150
329. Wurzburg OB, Jarowenko W, Patel JK, Rubens RW (1979) US 4166173 CAN 91:212879
330. Greidinger DS, Cohen BM (1967) US 3320237 CAN 67:91923
331. Park DP, Sung JH, Choi HJ, Jhon MS (2004) J Mater Sci 39:6083
332. Landerito NA, Wang Y-J (2005) Cereal Chem 82:264
333. Solarek DB (1987) Phosphorylated starches and miscellaneous inorganic esters. Wurzburg OB (ed) Modified Starches: Properties and uses. CRC Press, Boca Raton, pp 97
334. Lewandowicz G, Szymanska G, Voelkel E, Walkowski A (2000) Pol J Food Nutr Sci 9:31
335. Rutenberg MW (1980) Starch and its modification. In: Davidson RL (ed) Handbook of Water Soluble Gums and Resins. McGraw-Hill, New York, Chap 22
336. Kerr RW, Cleveland FC Jr (1959) US 2884413 CAN 53:91964
337. Swiderski F (1977) Acta Aliment Pol 3:115
338. Hjermstad ET (1962) US 3069411 CAN 58:34158
339. Christoffel C, Borel EA, Blumenthal A, Schobinger U, Mueller K (1967) US 3352848
340. Park DP, Sung JH, Kim CA, Choi HJ, Jhon MS (2004) J Appl Polym Sci 91:1770
341. Waly A, Abdel-Mohdy FA, Higazy A, Hebeish A (1994) Starch/Staerke 46:59
342. Sannella JL, Whistler RL (1963) Arch Biochem Biophys 102:226
343. Whistler RL (1969) Denpun Kogyo Gakkaishi 17:41
344. Towle GA, Whistler RL (1972) Phosphorylation of starch and cellulose with an amine salt of tetra polyphosphoric acid. In: Whistler RL, BeMiller JN (eds) Methods in Carbohydrate Chemistry, vol 6. Academic Press, New York, pp 408
345. Kajikawa M, Santo H (1998) JP 10152503 CAN 129:82948
346. Yeh MH (1992) EP 510986 CAN 118:171406
347. Sato T, Nishimura-Uemura J, Shimosato T, Kawai Y, Kitazawa H, Saito T (2004) J Food Prot 67:1719
348. Nishi N, Ebina A, Nishimura S, Tsutsumi A, Hasegawa O, Tokura S (1986) Int J Biol Macromol 8:311
349. Nishi N, Maekita Y, Nishimura S, Hasegawa O, Tokura S (1987) Int J Biol Macromol 9:109
350. Dace R, McBride E, Brooks K, Gander J, Buszko M, Doctor VM (1997) Thromb Res 87:113
351. Touey GP (1956) US 2759924 CAN 51:3440

352. Nehls I, Loth F (1991) Acta Polym 42:233
353. Ahn B-G (2003) Polym J (Tokyo) 35:23
354. Granja PL, Pouysegu L, Petraud M, De Jeso B, Baquey C, Barbosa MA (2001) J Appl Polym Sci 82:3341
355. Vigo TL, Welch CM (1973) Textilveredlung 8:93
356. Wagenknecht W, Philipp B, Schleicher H (1979) Acta Polym 30:108
357. Klemm D, Philip B, Heinze T, Heinze U, Wagenknecht W (1998) Comprehensive Cellulose Chemistry, vol 2. Wiley, New York, pp 136
358. Wagenknecht W, Philipp B, Nehls I, Schnabelrauch M, Klemm D, Hartmann M (1991) Acta Polym 42:213
359. Wagenknecht W, Nehls I, Philipp B, Schnabelrauch M, Klemm D, Hartmann M (1991) Acta Polym 42:554
360. Wagenknecht W (1996) Papier (Bingen, Germany) 50:712
361. Crater W de C (1933) US 1922123 CAN 27:57762
362. Mustafa A, Dawoud AF, Marawan A (1970) Starch/Staerke 22:17
363. Wolfrom ML, Maher GG, Chaney A (1958) J Org Chem 23:1990
364. Caesar GV (1958) Starch nitrate. In: Wolfrom ML, Tipson RS (eds) Advances in Carbohydrate Chemistry, vol 13. Academic Press, New York, pp 331
365. Brewer RJ, Bogan RT (1985) Cellulose esters, inorganic. In: Mark HF, Bikales NM, Overberger CG, Menges G (eds) Encyclopedia of Polymer Science and Engineering, vol 3. Wiley, New York, pp 139
366. Alexander WJ, Mitchell RL (1949) Anal Chem 21:1497
367. Bennett CF, Timell TE (1955) Sven Papperstidn 58:281
368. Thinius K, Thümler W. (1966) Makromol Chem 99:117
369. Clermont LP, Bender F (1972) J Polym Sci Part A. Polym Chem 10:1669
370. Schweiger RG (1977) US 4035569
371. Green JW (1963) Nitration with mixtures of nitric and sulfuric acids. In: Whistler RL (ed) Methods in Carbohydrate Chemistry, vol III. Academic Press, New York, pp 213
372. Eliseeva LM, Berenshtein EI, Aikhodzhaev BI, Pogosov YL (1968) Zh Anal Khim 23:436
373. Malm CJ, Genung LB, Williams RF Jr, Pile MA (1944) Ind Eng Chem Anal Ed 16:501
374. Malm CJ, Nadeau GF (1937) US 2069892 CAN 31:17395
375. Yang CQ, Wang D (2000) Text Res J 70:615
376. Hu X, Zhou X (1998) J China Text Univ (Engl Ed) 15:10
377. Shkol'nik SI, Teodorovich DA, Chernaya MA (1973) Issled Razrab Poligr Prom 64
378. Abo-Shosha MH, Ibrahim NA, Elnagdy EI, Gaffar MA (2002) Polym Plast Technol Eng 41:963
379. Proinov D, Georgiev G (1993) God Sofii Univ "Sv Kliment Okhridski" Khim Fak 86:83
380. Prey V, Schindlbauer H, Maday E (1973) Starch/Staerke 25:73
381. Ogawa K, Hirai I, Shimasaki C, Yoshimura T, Ono S, Rengakuji S, Nakamura Y, Yamazaki I (1999) Bull Chem Soc Jpn 72:2785
382. Krasovskii AN, Polyakov DN, Mnatsakanov SS (1993) J Appl Chem 66:1118
383. Krasovskii AN, Plodistyi AB, Polyakov DN (1996) J Appl Chem 69:1183
384. Sollinger S, Diamantoglou M (1996) Papier (Bingen, German) 50:691
385. Chong CK, Xing J, Phillips DL, Corke H (2001) J Agric Food Chem 49:2702
386. Malm CJ, Genung LB, Kuchmy W (1953) Anal Chem 25:245
387. Klaushofer H, Berghofer E, Pieber R (1979) Starch/Staerke 31:259
388. Goodlett VW, Dougherty JF, Patton HW (1971) J Polym Sci Part A: Polym Chem 9:155
389. Kamide K, Okajima K (1981) Polym J (Tokyo) 13:127
390. Buchanan CM, Hyatt JA, Lowman DW (1987) Macromolecules 20:2750
391. Gagnaire DY, Taravel FR, Vignon MR (1976) Carbohydr Res 51:157
392. Gagnaire DY, Taravel FR, Vignon MR (1982) Macromolecules 15:126
393. Capon B, Rycroft DS, Thomson JW (1970) Carbohydr Res 70:145
394. Miyamoto T, Sato Y, Shibata T, Inagaki H, Tanahashi M (1984) J Polym Sci Polym Chem Ed 22:2363
395. Sei T, Ishitani K, Suzuki R, Ikematsu K (1985) Polym J (Tokyo) 17:1065
396. Kamide K, Okajima K, Kowsaka K, Matsui T (1987) Polym J (Tokyo) 19:1405
397. Kowsaka K, Okajima K, Kamide K (1986) Polym J (Tokyo) 18:827
398. Kamide K, Saito M (1994) Macromol Symp 83:233
399. Buchanan CM, Edgar KJ, Hyatt JA, Wilson AK (1991) Macromolecules 24:3050
400. Hikichi K, Kakuta Y, Katoh T (1995) Polym J (Tokyo) 27:659
401. Tezuka Y, Tsuchiya Y (1995) Carbohydr Res 273:83
402. Lee CK, Gray GR (1995) Carbohydr Res 269:167; (a) Yu N, Gray RG (1998) Carbohydr Res 312:29

403. Iwata T, Azuma J, Okamura K, Muramoto M, Chun B (1992) Carbohydr Res 224:277
404. Hornig S (2005) Selbststrukturierende Funktionspolymere durch chemische Modifizierung von Dextranen. Diploma Thesis, University of Jena
405. Hussain MA, Liebert T, Heinze T (2004) Macromol Rapid Commun 25:916
406. Buchanan CM, Hyatt JA, Lowman DW (1989) J Am Chem Soc 111:7312
407. Nunes T, Burrows HD, Bastos M, Feio G, Gil MH (1995) Polymer 36:479
408. Iijima H, Kowsaka K, Kamide K (1992) Polym J (Tokyo) 24:1077
409. Mischnick P, Heinrich J, Gohdes M, Wilke O, Rogmann N (2000) Macromol Chem Phys 201:1985
410. DeBelder AN, Norrman B (1968) Carbohydr Res 8:1
411. Bouveng HO (1961) Acta Chem Scand 15:78
412. Liebert T, Pfeiffer K, Heinze T (2005) Macromol Symp 223:93
413. Bjorndal H, Lindberg B, Rosell KG (1971) J Polym Sci Part C. Polym Symp 36:523
414. Prehm P (1980) Carbohydr Res 78:372
415. Mischnick P (1991) J Carbohydr Chem 10:711
416. Gohdes M, Mischnick P, Wagenknecht W (1997) Carbohydr Polym 33:163
417. Garegg PJ, Lindberg B, Konradsson P, Kvarnstrom I (1988) Carbohydr Res 176:145
418. Stevenson TT, Furneaux RH (1991) Carbohydr Res 210:277
419. Mischnick-Luebbecke P, Koenig WA, Radeloff M (1987) Starch/Staerke 39:425
420. Heinrich J, Mischnick P (1999) J Polym Sci Polym Chem Ed 37:3011
421. Borjihan G, Zhong GY, Baigude H, Nakashima H, Uryu T (2003) Polym Adv Technol 14:326
422. Demleitner S, Kraus J, Franz G (1992) Carbohydr Res 226:239
423. Hirano S, Ohe Y, Ono H (1976) Carbohydr Res 47:315
424. Hirano S, Moriyasu T (1981) Carbohydr Res 92:323
425. Malm CJ, Tanghe LT, Laird BC, Smith GD (1952) J Am Chem Soc 74:4105
426. Malm CJ, Barkey KT, Salo M, May DC (1957) J Ind Eng Chem 49:79
427. Kowsaka K, Okajima K, Kamide K (1988) Polym J (Tokyo) 20:827
428. Philipp B, Wagenknecht W, Wagenknecht M, Nehls I, Klemm D, Stein A, Heinze Th, Heinze U, Helbig K (1995) Papier (Bingen, Germany) 49:3
429. Klemm D, Heinze T, Stein A, Liebert T (1995) Macromol Symp 99:129
430. Altaner C, Saake B, Puls J (2003) Cellulose (Dordrecht, Netherlands) 10:85
431. Puls J, Altaner C, Saake B (2004) Macromol Symp 208:239
432. Altaner C, Saake B, Puls J (2003) Cellulose (Dordrecht, Netherlands) 10:391
433. Camacho Gomez JA, Erler UW, Klemm DO (1996) Macromol Chem Phys 197:953
434. Isogai A, Ishizu A, Nakano J (1986) J Appl Polym Sci 31:341
435. Iwata T, Doi Y, Azuma J (1997) Macromolecules 30:6683
436. Tsunashina Y, Hattori K (2000) J Colloid Interface Sci 228:279
437. Tsunashima Y, Hattori Kawanishi H, Hori F (2001) Biomacromolecules 2:991
438. Klemm D, Heinze T, Philipp B, Wagenknecht W (1997) Acta Polym 48:277
439. Heinze T, Vieira M, Heinze U (2000) Lenzinger Ber 79:39
440. Hatanaka K, Tomioka N, Ohta S, Kadokura T, Zulueta Kasuya MC (1996) Nippon Setchaku Gakkaishi 32:256
441. Heinze T, Heinze U, Grote C, Kötz J, Lazik W (2001) Starch/Staerke 53:261
442. Xie J, Hsieh YL (2001) Polym Prepr (Am Chem Soc Div Polym Chem) 42:512
443. Einfeldt L, Petzold K, Günther W, Stein A, Kussler M, Klemm D (2001) Macromol Biosci 1:341
444. Klemm D, Stein A, Erler U, Wagenknecht W, Nehls I, Philipp B (1993) New procedures for regioselective synthesis and modification of trialkylsilylcelluloses. In: Kennedy JF, Phillips GO, Williams PA (eds) Cellulosics: Materials for Selective Separation and Other Technologies. Ellis Horwood, New York, pp 221
445. Petzold K, Koschella A, Klemm D, Heublein B (2003) Cellulose (Dordrecht, Netherlands) 10:251
446. Koschella A, Heinze T, Klemm D (2001) Macromol Biosci 1:49
447. Edgar KJ, Buchanan CM, Debenham JS, Rundquist PA, Seiler BD, Shelton MC, Tindall D (2001) Prog Polym Sci 26:1605
448. Glasser WG (2004) Macromol Symp 208:371
449. Staude E (1992) Membranen und Membranprozesse. Wiley, Weinheim
450. Okuyama K, Obata Y, Noguchi K, Kusaba T, Ito Y, Ohno S (1996) Biopolymers 38:557
451. Shao B-H, Xu X-Z, Wu Q-Z, Lu J-D, Fu X-Y (2005) J Liq Chromatogr Relat Technol 28:63
452. Sei T, Matsui H, Shibata T, Abe S (1992) ACS Symp Ser 489:53

453. Kubota T, Yamamoto C, Okamoto Y (2004) J Poly Sci Polym Chem Ed 42:4704
454. Bauer R, Breu W, Wagner H, Weigand W (1991) J Chromatogr 541:464
455. Stampfli H, Patil G, Sato R, Quon CY (1990) J Liq Chromatogr 13:1285
456. Ficarra R, Calabro ML, Tommasini S, Grasso S, Carulli M, Monforte AM, Costantino D (1995) Chromatographia 40:39
457. Garces J, Franco P, Oliveros L, Minguillon C (2003) Tetrahedron. Asymmetry 14:1179
458. Han X, Wen X, Guan Y, Zhao Li, Li C, Chen L, Li Y (2004) Fenxi Huaxue 32:1287
459. Okamoto Y, Yajima E, Yamamoto T (2002) JP 2002148247 CAN 136:395058
460. Kasuya N, Nakashima J, Kubo T, Sawatari A, Habu N (2000) Chirality 12:670
461. Nishimura M, Hashida I, Minamii N, Sugiyama H, Iwamoto O (1986) JP 61153101 CAN 106:121739
462. Dautzenberg H, Schuldt U, Lerche D, Woehlecke H, Ehwald R (1999) J Membr Sci 162:165
463. Dautzenberg H, Lukanoff B, Eckert U, Tiersch B, Schuldt U (1996) Ber Bunsen-Ges 100:1045
464. Zhang J, Yao S-J, Guan Y-X (2005) J Membr Sci 255:89
465. Karle P, Mueller P, Renz R, Jesnowski R, Saller R, Von Rombs K, Nizze H, Liebe S, Gunzburg WH, Salmons B, Lohr M (1998) Adv Exp Med Bio 451:97
466. Kammertoens T, Gelbmann W, Karle P, Alton K, Saller R, Salmons B, Gunzburg WH, Uckert W (2000) Cancer Gene Ther 7:629
467. Motomura T, Miyashita Y, Ohwada T, Onishi M, Yamamoto N (1995) EP 679436 CAN 124:37673
468. Izumi J, Yoshikawa M, Kitao T (1997) Maku 22:149
469. Yoshikawa M, Ooi T, Izumi J-I (1999) J Appl Polym Sci 72:493
470. Ramamoorthy M, Ulbricht M (2003) J Membr Sci 217:207
471. Ulbricht M, Malaisamy R (2005) J Mater Chem 15:1487
472. Franz G, Alban S (1995) Int J Biol Macromol 17:311
473. Kamitakahara H, Nishigaki F, Mikawa Y, Hori M, Tsujihata S, Fujii T, Nakatsubo F (2002) J Wood Sci 48:204
474. Yoshida T, Hatanaka K, Uryu T, Kaneko Y, Suzuki E, Miyano H, Mimura T, Yoshida O, Yamamoto N (1990) Macromolecules 23:3717
475. Havlik I, Looareesuwan S, Vannaphan S, Wilairatana P, Krudsood S, Thuma PE, Kozbor D, Watanabe N, Kaneko Y (2005) Trans R Soc Trop Med Hyg 99:333
476. Kindness G, Long WF, Williamson FB (1979) IRCS Med Sci: Libr Compend 7:134
477. Kamide K, Okajima K, Matsui T, Kobayashi H (1984) Polym J (Tokyo) 16:259
478. Escartin Q, Lallam-Laroye C, Baroukh B, Morvan FO, Caruelle JP, Godeau G, Barritault D, Saffar JL (2003) FASEB J 17:644
479. Uryu T (2001) Kikan Kagaku Sosetsu 48:103
480. Doctor VM, Lewis D, Coleman M, Kemp MT, Marbley E, Sauls V (1991) Thromb Res 64:413
481. Mori S, Sugawara I, Ito W (1989) EP 342544 CAN 113:84814
482. Li M, Huang P, Kong X (2004) Xiandai Zhongliu Yixue 12:532
483. Vongchan P, Sajomsang W, Subyen D, Kongtawelert P (2002) Carbohydr Res 337:1239
484. Suzuki M, Mikami T, Matsumoto T, Suzuki S (1977) Carbohydr Res 53:223
485. Di Benedetto M, Starzec A, Colombo BM, Briane D, Perret GY, Kraemer M, Crepin M (2002) Br J Pharmacol 135:1859
486. Wang X, Feng Q, Cui F (2004) CN1470247 CAN 142:170128
487. Miyatake K, Okamoto Y, Shigemasa Y, Tokura S, Minami S (2003) Carbohydr Polym 53:417
488. Kamide K, Okajima K, Matsui T, Ohnishi M, Kobayashi H (1983) Polym J (Tokyo) 15:309
489. Mizumoto Kenji, Sugawara I, Ito W, Kodama T, Hayami M, Mori S (1988) Jpn J Exp Med 58:145
490. Gao Y, Katsuraya K, Kaneko Y, Mimura T, Nakashima H, Uryu T (1998) Polym J (Tokyo) 30:31
491. Gao Y, Katsuraya K, Kaneko Y, Mimura T, Nakashima H, Uryu T (1999) Macromolecules 32:8319
492. Dang W, Colvin OM, Brem H, Saltzman WM (1994) Cancer Res 54:1729
493. Larsen C (1986) Acta Pharm Suec 23:279
494. Larsen C, Kurtzhals P, Johansen M (1988) Acta Pharm Suec 25:1
495. Lee JS, Jung YJ, Doh MJ, Kim YM (2001) Drug Dev Ind Pharm 27:331
496. Jung YJ, Lee JS, Kim HH, Kim YT, Kim YM 1998) Arch Pharmacal Res 21:179
497. Jung YJ, Lee JS, Kim HH, Kim YT, Kim YM, Kim DD, Han SK (1998) Yakhak Hoechi 42:31
498. Larsen C (1989) J Pharm Biomed Anal 7:1173
499. Harboe E, Johansen M, Larsen C (1988) Farm Sci Ed 16:73
500. Larsen C, Jensen BH, Olesen HP (1991) Acta Pharm Nord 3:71

501. Larsen C, Johansen M (1989) Acta Pharm Nord 1:57
502. Sungwon Kb, Su YC, Kun N, Sung WK, You HB (2003) Biomaterials 24:4843
503. Kreuter J (1991) J Controlled Release 16:169
504. Gref R, Minamitake Y, Peracchia MT, Trubetskoy V, Torchillin V, Langer R (1994) Science (Washington, DC) 263:1600
505. Jeong Y-I, Cheon JB, Kim S-H, Nah JW, Lee YM, Sung YK, Akaite T, Cho CS (1998) J Controlled Release 51:169
506. Liebert T, Hornig S, Hesse S, Heinze T (2005) J Am Chem Soc 127:10484
507. Lee KY, Jo WH, Kwon IC, Kim YH, Jung SW (1998) Macromolecules 31:378
508. Kim S-H, Won C-Y, Chu C-C (2000) WO 2000012619 CAN 132:208882
509. Kim S-H, Won C-Y, Chu C-C (1999) J Biomed Mater Res 46:160
510. Yamaoka T, Tanihara M, Mikami H, Kinoshita H (2003) JP 2003252936 CAN 139:235488
511. Adams HA (1991) Anaesthesiol Intensivmed 10:277
512. Warnken UH, Asskali F, Förster H (2000) Starch/Staerke 52:261
513. Warnken UH, Asskali F, Förster H (2001) Krankenhauspharmazie 22:113
514. Heins D, Kulicke W-M, Käuper P, Thielking H (1998) Starch/Staerke 50:431
515. Hofmann R (1929) DE 526479 CAN 25:39014
516. Bates H, Fisher JW, Smith JR (1955) GB 737566 CAN 50:38373
517. Aburto J, Thiebaud S, Alric I, Borredon E, Bikiaris D, Prinos J, Panayiotou C (1997) Carbohydr Polym 34:101
518. Edgar KJ, Bogan RT (1996) WO 9620960 CAN 125:198904
519. Jeanes AR, Wilham CA (1952) US 2587623 CAN 46:54692
520. Stadler PA (1978) Helv Chim Acta 61:1675
521. Klemm D, Schnabelrauch M, Stein A, Philipp B, Wagenknecht W, Nehls I (1990) Papier 44:624
522. Tiefenthaler KHO, Wyss U (1981) EP 30443 CAN 95:117405
523. Lohmar R, Sloan JW, Rist CE (1950) J Am Chem Soc 72:5717
524. Koschella A, Klemm D (1997) Macromol Symp 120:115
525. Klemm D, Philipp B, Heinze T, Heinze U, Wagenknecht W (1998) Comprehensive Cellulose Chemistry, vol II. Wiley, New York, pp 235

Subject Index